GEOCHEMISTRY RESEARCH ADVANCES

GEOCHEMISTRY RESEARCH ADVANCES

ÓLAFUR STEFÁNSSON
EDITOR

Nova Science Publishers, Inc.
New York

Copyright © 2008 by Nova Science Publishers, Inc.

All rights reserved. No part of this book may be reproduced, stored in a retrieval system or transmitted in any form or by any means: electronic, electrostatic, magnetic, tape, mechanical photocopying, recording or otherwise without the written permission of the Publisher.

For permission to use material from this book please contact us:
Telephone 631-231-7269; Fax 631-231-8175
Web Site: http://www.novapublishers.com

NOTICE TO THE READER

The Publisher has taken reasonable care in the preparation of this book, but makes no expressed or implied warranty of any kind and assumes no responsibility for any errors or omissions. No liability is assumed for incidental or consequential damages in connection with or arising out of information contained in this book. The Publisher shall not be liable for any special, consequential, or exemplary damages resulting, in whole or in part, from the readers' use of, or reliance upon, this material. Any parts of this book based on government reports are so indicated and copyright is claimed for those parts to the extent applicable to compilations of such works.

Independent verification should be sought for any data, advice or recommendations contained in this book. In addition, no responsibility is assumed by the publisher for any injury and/or damage to persons or property arising from any methods, products, instructions, ideas or otherwise contained in this publication.

This publication is designed to provide accurate and authoritative information with regard to the subject matter covered herein. It is sold with the clear understanding that the Publisher is not engaged in rendering legal or any other professional services. If legal or any other expert assistance is required, the services of a competent person should be sought. FROM A DECLARATION OF PARTICIPANTS JOINTLY ADOPTED BY A COMMITTEE OF THE AMERICAN BAR ASSOCIATION AND A COMMITTEE OF PUBLISHERS.

LIBRARY OF CONGRESS CATALOGING-IN-PUBLICATION DATA

Geochemistry research advances / Olafur Stefánsson (editor).
   p. cm.
 ISBN 978-1-60456-215-6 (hardcover)
 1. Geochemistry--Research. I. Stefánsson, Olafur, 1959-
QE515.G4145 2008
551.9--dc22
                     2007049761

Published by Nova Science Publishers, Inc. ✦ *New York*

CONTENTS

Preface		vii
Chapter 1	Implications of Complex Glacial Deposits for Till Geochemical Exploration: Examples from the Central Fennoscandian Ice Sheet *P. Sarala and V.J Ojala*	1
Chapter 2	Geochemical Modeling of Concentrated Mine Waters: A Comparison of the Pitzer Ion-Interaction Theory with the Ion-Association Model for the Study of Melanterite Solubility in San Telmo Mine (Huelva, Spain) *Javier Sánchez España and Marta Diez Ercilla*	31
Chapter 3	Geochemical Anomalies Connected with Great Earthquakes in China *Jianguo Du, Xueyun Si, Yuxiang Chen, Hong Fu, Chunlin Jian and Wensheng Guo*	57
Chapter 4	Structural Characterization of Kerogen by Ruthenium Tetroxide Oxidation *Veljko Dragojlovic*	93
Chapter 5	Mucous Macroaggregates in the Northern Adriatic *Nives Kovač, Jadran Faganeli and Oliver Bajt*	119
Chapter 6	Geochemical Signals and Paleoclimate Changes in a 16,000 ^{14}C Year Sedimentary Record from Lake Gucheng, Eastern China *R.L. Wang, S.C. Zhang, S.C. Brassell, D. Tomasi, S.C. Scarpitta, G. Zhang, G.Y. Sheng, S.M. Wang, and J.M. Fu*	143
Chapter 7	Biogeochemical Evaluation of Soil Covers for Base Metal Tailings, Ag-Pb-Zn Cannington Mine, Australia *Benjamin S. Gilfedder and Bernd G. Lottermoser*	163
Chapter 8	The Geochemical Characteristics of Rare Earth Elements in Granitic Laterites in Hainan Island, China *Liu Qiang, Bi Hua, Wang Xueping, Wang Minying, Zhao Zhizhong, Yang Yuangen and Zhu Weihuang*	181
Index		191

PREFACE

The field of geochemistry involves study of the chemical composition of the Earth and other planets, chemical processes and reactions that govern the composition of rocks and soils, and the cycles of matter and energy that transport the Earth's chemical components in time and space, and their interaction with the hydrosphere and the atmosphere. This new book presents leading research in the field.

Chapter 1 - Till sampling in different types glacigenic deposits and the interpretation of results are very challenging exercise. The experience from northern Finland shows that the composition and stratigraphy of till cover and the specific characteristics of separate moraine formations must be carefully considered to success in exploration. Finnish Lapland was repeatedly in the centre of continental glaciers with related strong changes in glacial dynamics, ice lobation and ice-flow directions. In this environment rapid changes between erosional and depositional conditions have led to variable degree of preservation of earlier deposits and pre-Quaternary regolith, and deposition of complex glacigenic formations.

Till geochemistry with surficial geology are used as a practical exploration tool in glaciated areas since 1950's, increased knowledge and better understanding of glacial deposits have given an opportunity to direct exploration and, also re-examine older results and datasets in a new way. Furthermore, a multidisplinary approach using various exploration methods allows better interpretation and analysis of the results.

In this article the authors will present how the latest surficial geological, till geochemical and heavy mineral studies in northern Finland have improved the estimation of glacial transport mechanisms, dispersion, directions and distances in the central part of the Fennoscandian glaciations. Many recent and on-going studies have shown that specifically the identification and classification of surficial deposits and determination of the formation mechanisms of glacial deposits needs to be integrated into exploration as early as possible.

Chapter 2 - This work describes recent hydrogeochemical research carried out in an extremely acidic (pH 0.61-0.82) hypersaline brine (TDS=225 g/L, including 74 g/L (1.33 M) Fe, 134 g/L (1.39 M) SO_4^{2-}, 7.5 g/L (0.28 M) Al, 3 g/L (0.12 M) Mg, 2 g/L (0.03 M) Cu, and 1 g/L (0.015 M) Zn) which seeps from a pyrite pile and forms evaporative pools of green, mostly ferrous, ultra-concentrated water in the massive sulphide mine of San Telmo (Huelva, Iberian Pyrite Belt, SW Spain). The physico-chemical conditions of this water allow the formation of attractive crystals of Zn-rich melanterite ($Fe^{II}SO_4 \cdot 7H_2O$), whose solubility is related with the concentration of Fe(II) and SO_4^{2-}, as well with pH and T. Geochemical modeling carried out with the PHREEQCI computer code using the ion-association model with the Davis equation and the Pitzer specific-ion-interaction approach with MacInnes convention, indicates that, although both models have limitations, the Pitzer approach describes better the water/melanterite equilibrium problem. The Davis approach can not explain the formation of this mineral at the conditions found in San Telmo (SI_{Mel}=-0.90 to -0.50), whereas the Pitzer model suggests that the acidic brine was very slightly subsaturated with respect to melanterite (SI_{Mel}=0.0 to -0.40) and predicts melanterite precipitation at cool temperatures (T<2-3 °C) and/or Fe and SO_4^{2-} concentrations slightly higher than the ones measured in the water, which may perfectly account for the presence of this mineral in the studied system. Geochemical research using advanced modeling tools may help to solve water/mineral equilibrium problems in solutions of extreme ionic strength such as ultra-concentrated mine waters.

Chapter 3 - The goals of this chapter are to investigate the relationship between the geochemical anomalies and great earthquakes and to distinguish seismic precursors from the abnormal phenomena in order to improve the accuracy of predicting an impending earthquake. Many devastating earthquakes occurred in China. The great earthquakes left many deaths and caused a lot of economic loss. Earthquake prediction is considered as one of the most efficient approaches to mitigate seismic hazard. Predicting earthquake mainly depends on understanding the process of earthquake generation and mechanism of seismic precursors.

The establishment and development of the seismic monitoring network in China are introduced. Plenty of novel geochemical data in China has been obtained since the late 60's of last century. The geochemical anomalies frequently occurred in the seismic zones, but rarely correlated with great earthquakes. Geochemical anomalies related to the 3 June 2007 Puer M6.4 earthquake were described. The temporal and spatial features of geochemical anomalies connected with great earthquakes varied dramatically. Combining with the geological heterogeneousness that is not well understood, it is difficult to put forward the methods to correctly identify the seismic precursors. There are a few of successful examples for predicting earthquakes based on the gaseous and hydrochemical anomalies and other precursors, but a lot of failure ones.

The geochemical anomalies related to great earthquakes can be attributed to fluid mixing and water-rock interaction. The experimental data for simulating water-rock interaction associated with formation of micro-fracture in a brittle aquifer demonstrate the two-steps model: (1) water mixed with fluids of fluid inclusions in rock resulted in soluble Cl and SO_4 approached approximately equilibrium in six hours or less; (2) dissolution predominantly controlled concentrations of other ions, resulting in the concentrations increased with increasing soaking time. A genetic model, in which the role of fluids is emphasized, is proposed based on geological and geophysical investigations and experiments of rocks at high pressure and high temperature in order to highlight the mechanism of the seismic-geochemical anomalies and process of earthquake gestation.

Chapter 4 - Kerogen is macromolecular sedimentary organic substance found in oil sands and oil shales. Kerogens of various types have been extensively studied by oxidative degradation methods. Main difficulties in reconstruction of the kerogen macromolecular structure, based on the products of an oxidative degradation, result from lack of information about the original structural elements. Over the past twenty years, ruthenium tetroxide oxidation, in combination with other analytical methods, has been extensively used in structural elucidation and has provided considerable insight into structure and origin of various kerogens. A somewhat unique feature of ruthenium tetroxide is that, although it is highly reactive, it is also a selective oxidizing agent. It is more selective compared to other common oxidants such as permanganate, chromate or ozone. Ruthenium tetroxide oxidizes alcohols, aldehydes, alkenes and alkynes to the corresponding carboxylic acids. Compared to the chromate or ozone oxidation, yields of ruthenium tetroxide oxidation are considerably better, which allows for a greater confidence in interpreting the results. While permanganate oxidation proceeds in similar yields, ruthenium tetroxide oxidation is considerably faster and simpler method. Furthermore, in some cases its chemoselectivity differs from that of permanganate and one can obtain additional information by combining the results of the two methods. A number of research groups used ruthenium tetroxide degradation to study composition of the aliphatic portion of the sedimentary organic matter. Nevertheless, in some cases, it is possible to obtain high yields of aromatic products. One of its most useful and unique features of ruthenium tetroxide is that it oxidizes ethers to the corresponding esters, thus allowing differentiation of ether moieties from esters. It has led to an extensive study of extant and fossil algaenans, which culminated into development of a method for their chemical fingerprinting.

Chapter 5 - The episodic hyperproduction of mucous macroaggregates in the northern Adriatic, offers a rare opportunity to study the assembling of macromolecular DOM into macrogels and macroaggregates. They represent an important site of accumulation, transformation and degradation of organic matter, contributing to the patchiness, distribution and fate of particulate matter in seawater. In this chapter the results of our research work of several years is presented and combined with the results of some other authors. To elucidate this phenomenon, their biological and chemical composition, formation and degradation processes and finally their environmental role are presented and discussed. Emphasis is given to the use of spectroscopic techniques, e.g. ^{1}H NMR, ^{13}C NMR and FTIR, which are usually used for the determination of the chemical composition of organic compounds, to decoding the macroaggregate composition and structure.

Chapter 6 - A long sedimentary core (20 m) taken from Lake Gucheng, Nanjing, Jiangsu Province of eastern China was studied focusing on geochemical changes versus paleoclimatic and hydrological history. ^{14}C dating results of fossil organic carbon from the GS-1 core indicate the oldest sample was deposited around 16 kyr BP. The location of this lake at the intersection of monsoon climate regions in continental China provides a valuable opportunity to study the Late Quaternary climate history. Organic and inorganic carbon contents, stable carbon and oxygen isotopic compositions (δ^{13}C and d^{18}O values), trace metal elements, coupled with indicative mineral contents, provide detailed comprehensive geochemical signals suggesting marked climate and hydrological changes in this region since the end of last glacial maximum. Sediment samples from the lower part (19.7 – 12.5 m; 15.9 – 10 kyr BP) of the cores are carbon, trace metal elements (Cr, Co, Zn, Pb, Ni etc.) and Na/Fe ratios are lower than those of the upper part of the cores (12.5 – 0 m; 10 kyr BP - recent), suggesting a deeper lake environment developed following the end of the Last Glacial maximum and the enhanced monsoon precipitation and warming up in the region starting at the beginning of Holocene. Stable isotope data, however, shows more fluctuations rather than the two clearly cut stages between the Late Pleistocene and the Holocene as shown by the mineral, elemental and lithological data. These isotope data show that several important paleoclimatic events including the ending of the Last Glacial Maximum (16 - 15 kyrBP), the severe cold/dry event Younger Dryas (10 - 11 kyrBP), the optimum climate during the mid-Holocene, and a substantial cooling and drought episode during the Iron Age Neoglaciation in later Holocene (3500 - 2000 yrBP), may all be observed with geochemical signals in this inland lake despite the overlapping complications of the other no-climatically induced geochemistry effects such as sediment diagenesis and hydrological environment changes.

Chapter 7 - This study reports on the transfer of metals from soil covered tailings into native plants at the Cannington Ag-Pb-Zn mine in semi-arid northwest Queensland, Australia. A number of field trial plots were established over sulfidic metal-rich tailings in 2001. The plots differed in either soil depth (200, 500, 800 mm) or the combination of local soil and waste rock (1600 mm) used for the construction of the trialed capping strategy. In 2004, the plots were sampled for their cover materials, cover plants and tailings to evaluate the performance of the different cover designs. In all field trial plots, the roots and to a lesser degree the foliage of native plant species (Triodia longiceps, Astrebla lappacea, Astrebla squarrosa, Iseilema membranaceum, Rhynchosia minima, Sclerolaena muricata), growing on the soil covered tailings, display evidence of biological uptake of Ag, As, Cd, Pb, Sb and Zn, with values being up to one order of magnitude above background samples for the same species. The plants acquired their detected metal distributions from the tailings and mineralized waste rocks as evidenced by the penetration of plant roots through the entire soil cover to the top of the tailings or the mineralized waste rock layer. In general, the plant species growing on the soil covered tailings have bioconcentration factors (BCF, metal concentration ratio in plant roots to DTPA-extractable soil) and translocation factors (TF, metal concentration ratio of plant foliage to roots) for As, Cd, Cu and Zn and for Ag and Pb greater than one, respectively.

The trialled covers allow the translocation and accumulation of trace metals into the above-ground biomass of metal-tolerant native plants. Such processes may introduce metals and metalloids into surrounding ecosystems despite the waste remaining physically isolated. Hence, engineered dry covers of mine waste repositories need to consider the root penetration depth of native plants as well as the bioavailability of metals and their possible translocation and accumulation into the above-ground tissue of cover plants.

Chapter 8 - In granitic laterites across Hainan Island the average grosses of REEs, light REEs and available REEs are 459.25, 386.99 and 199.06 μg/g respectively. The above three values are higher than that of the crust 186 μg/g, but the average gross of heavy REEs is 72.26 μg/g. Light REEs are enriched. Available REEs are high and have significantly positive correlation with gross of REEs, accounting for the total REEs in the range of 17.99%-80.25% and with mean value of 43.35%. It is also related to high temperature and humidity, strong chemical weathering and acid to weak acid soil environment in tropical Hainan Island. Across Hainan Island the average grosses of REEs, light REEs and available REEs decline along with low mountain-hilly land → central mountain → tableland-coastal plain on the horizontal spatial dimension, while the average grosses of heavy REEs decline along with central mountain →lower mountain-hilly land→ tableland-coastal plain. The average grosses of REEs, and Light REEs decline from bottom to top in laterite (Horizon C → Horizon B → Horizon A), while heavy REEs and available REEs decline from bottom to top (Horizon C → Horizon B → Horizon A) gradually on the vertical spatial dimension. However in tableland areas heavy REEs and available REEs have similar vertical change patterns as REEs and Light REEs. All laterites show negative Eu-anomaly to weak-medium extent. It shows relative enrichment of Ce and positive Ce-anomaly in top layer (Horizon A) of granitic laterites in all geomorphic units and in lower altitude tableland-coastal plain, but on the vertical spatial dimension it shows the change from no anomaly to weak negative anomaly with the depth of the soils.

Before the authors project was carried out, there was no report about systematic study on rare earth elements (i.e. REEs) in laterites in Hainan Island[1]. Over 5 years' study, the authors surveyed the typical profiles of laterites on different mother materials, in different structural landform and with different maturity, collected, processed and analyzed many samples in a systematic way. The authors have conducted REE fertilizing experiments both in potted plants and in field growing plants. In this article the authors focus on the geochemical characteristics of REE in granitic laterites in Hainan Island.

In: Geochemistry Research Advances
Editor: Ólafur Stefánsson, pp. 1-29

ISBN 978-1-60456-215-6
© 2008 Nova Science Publishers, Inc

Chapter 1

IMPLICATIONS OF COMPLEX GLACIAL DEPOSITS FOR TILL GEOCHEMICAL EXPLORATION: EXAMPLES FROM THE CENTRAL FENNOSCANDIAN ICE SHEET

P. Sarala and V.J. Ojala
Geological Survey of Finland, P.O. Box 77, FIN-96101 Rovaniemi, FINLAND
E-mail: pertti.sarala@gtk.fi

ABSTRACT

Till sampling in different types glacigenic deposits and the interpretation of results are very challenging exercise. The experience from northern Finland shows that the composition and stratigraphy of till cover and the specific characteristics of separate moraine formations must be carefully considered to success in exploration. Finnish Lapland was repeatedly in the centre of continental glaciers with related strong changes in glacial dynamics, ice lobation and ice-flow directions. In this environment rapid changes between erosional and depositional conditions have led to variable degree of preservation of earlier deposits and pre-Quaternary regolith, and deposition of complex glacigenic formations.

Till geochemistry with surficial geology are used as a practical exploration tool in glaciated areas since 1950's, increased knowledge and better understanding of glacial deposits have given an opportunity to direct exploration and, also re-examine older results and datasets in a new way. Furthermore, a multidisplinary approach using various exploration methods allows better interpretation and analysis of the results.

In this article we will present how the latest surficial geological, till geochemical and heavy mineral studies in northern Finland have improved the estimation of glacial transport mechanisms, dispersion, directions and distances in the central part of the Fennoscandian glaciations. Many recent and on-going studies have shown that specifically the identification and classification of surficial deposits and determination of the formation mechanisms of glacial deposits needs to be integrated into exploration as early as possible.

Keywords: Glacial geology, glacigenic deposits, morphology, till geochemistry, heavy minerals, gold, exploration, prospectivity mapping, Quaternary, Lapland, Finland.

INTRODUCTION

Till is a mixture of material from fresh bedrock surface, weathered bedrock and older sediments which are glacially eroded, transported and deposited. In many parts of glaciated terrain the till cover dominates and the bedrock outcrops are rare. Due to glacigenic nature till debris and fragments are always some distance derived from the source(s) and have dispersed to the direction of ice-flow, giving also a larger and more homogenised indication of source than the bedrock itself (Shilts 1975). An intensity of secondary dispersion is influenced by many different factors which depend on the variations in, for example, geology and topography, and sub-glacial conditions during till deposition. Glacial dynamics is also one of the most important things, because it is affecting to whole ice mass and is a mixture of effects caused by the climate and other environment factors. Glacial morphology is a straight evidence of variation in glacial dynamics, and it reflects the deposition conditions under and front of the glacier's body.

Surficial geology has been used as a practical exploration tool in glaciated areas since the beginning of 2000 century (Sauramo 1924) but the geochemical methods have been in use since the 1950's (Kauranne 1958, Wennervirta 1968, Shilts 1972, Kujansuu 1976). The main methods are till geochemistry and heavy mineral investigation. These methods are based on the use of secondary dispersion of till. By till sampling the dispersion pattern and the extent can be estimated and sometimes bordered. Sampling grids are critical in this view and the use of them depends on the source or source areas that are under exploration. Regional geochemical mappings with sparse grids are useful when finding suitable provenances for exploration. A good example of that is the mapping project that was carried out in Finland in 1980's (Salminen 1995, Salminen & Tarvainen 1995). The geochemical atlas for the fine fraction of till (< 0.06 mm) was also produced based on the same database (Koljonen 1992). But when targeting the exploration the local scale with dense grids is needed. Till sampling can be done using different equipment like percussion drilling but the estimation of sampling depth must be based on the till stratigraphy. For that purpose test pit excavations and preliminary till and heavy mineral sampling are usually needed (Hirvas & Nenonen 1990, Sarala 2005a).

Many studies have proven that the good knowledge of till stratigraphy is needed for succeeding in exploration (Hirvas et al. 1977, Hirvas & Nenonen 1990, Sarala et al. 1998, Sarala 2005a). An identification of different till units based on their composition, structure, lithology, stone orientation and relation to other deposits is a key for till stratigraphy. It is also a way to recognise the mechanisms of the till deposition and glacial environment where the deposition happens or has happened (cf. Dreimanis 1990). In the central parts of continental glacier, like in northern Finland, weathered bedrock, older tills and other sediment units can be preserved under cold-based, sub-glacial conditions (Hättestrand 1997, Sarala 2005a) and/or in the ice-divide zone (Hirvas 1991). Glacigenic deposits are thick and multiple glacial advances are seen in the till stratigraphy. Glacial transportation and deposition are also very complex in those cases (Hirvas & Nenonen 1990).

Increased knowledge and better understanding of glacial deposits have given an opportunity to direct exploration and, also re-examine older results and datasets in a new way. The use of glacial morphology for reconstruction of glacial conditions and palaeoenvironment during the moraine formation (cf. Kleman & Borgström 1996, Kleman et al. 1997,

Hättestrand 1997) has given the opportunity to estimate glacial dispersion and transport distances in different parts of glaciated terrain (Sarala et al. 2007a). It is possible because the characteristics and the ways the different moraine types are deposited are nowadays quite well-known (cf. Aario 1990, Aario & Peuraniemi 1992, Menzies & Shilts 1996, Sarala 2005b, Sarala & Peuraniemi 2007). Reanalysis of older till samples, the use of Geoscientific information systems (GIS) and the analysis of datasets with new, GIS-based modelling software are very useful tools in exploration and give effective way to focus studies (Harris et al. 2006, Nykänen 2006, Nykänen & Salmirinne 2007, Nykänen et al. 2007, Sarala et al. 2007b).

The aim of this article is to present how the latest surficial geological, till geochemical and heavy mineral studies in northern Finland have improved the estimation of glacial transport mechanisms, dispersion, directions and distances in the central part of the Fennoscandian glaciations. The importance of the identification and classification of surficial deposits, determination of their formation mechanisms and the use of multiple datasets are proven by using several exploration examples from the northern Finland.

GEOLOGICAL SETTINGS

Bedrock

Northern Finland bedrock (Figure 1) can be divided into six main units from north to south: the Archaean Inari Area, the Paleoprotorozoic Lapland Granulite Belt (ca 1.9 Ga), Paleoproterozoic Central Lapland Area (2.5-1.9 Ga), Paleoproterozoic Central Lapland Granitoid Complex (1.9-1.8 Ga), Paleoproterzoic Peräpohja Schist Belt (2.0-2.3 Ga) and the Archaean Pudasjärvi Complex (3.5-2.7 Ga). The area comprises a significant portion of the northern part of the Archaean Karelian craton and it records a prolonged and episodic history of sedimentation, rifting and magmatism throughout the Palaeoproterozoic times.

Archaean Inari Area
The Archaean Inari area is composed of several gneiss complexes. At least some of these complexes were exposed to weathering between 2.45-2.33 Ga (Sturt et al. 1994).

Gneiss complexes are intruded by 1.95 Ga to 1.90 Ga Paleoproterozoic granodioritic to gabbroic intrusions (Kesola 1995; Korsman et al. 1997; as well as Neoproterozoic diabase dikes, Vuollo & Huhma 2005), and by the 1.79 Ga Vainospää Granite. An evolved arc environment without strong involvement of Archaean crust is implied for some plutonic rocks in the Inari Area (1.94–1.93 Ga plutonic rocks; Barling et al. 1997). The Archaean complexes are strongly migmatised and reworking associated with Paleoproterozoic melting is reported from 1.95 Ga to 1.93 Ga (Kesola 1991).

Lapland Granulite Belt
The Lapland Granulite Belt contains arc-related psammitic and pelitic metasediments (garnet-sillimanite gneisses), anorthosites as one major unit and as concordant sheets, and noritic to enderbitic intrusions (pyroxene granulites; Barbey et al. 1986; Bibikova et al. 1993; Korja et al. 1996). The presence of anorthosites and ultramafics within the Lapland Granulite

Belt is argued to provide evidence for mantle input (Korja et al. 1996). Lapland Granulite Belt rocks have experienced granulite-facies metamorphic conditions, divided by Marker (1988) into a lower high-strain migmatitic part and an upper migmatitic part that experienced lower strain. Conditions for the peak high temperature metamorphic stage vary from 750°C to 850°C and 5.0 to 8.5 kilobars, and approximately 650°C and 2-3 kilobars for the final low temperature stage (Tuisku et al. 2006).

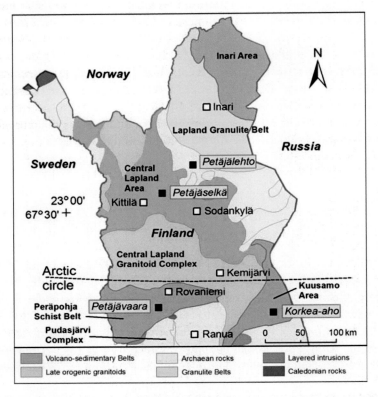

Figure 1: Bedrock map of northern Finland based on geological map of Finland 1:1 000 000. The main target areas are named with grey background. Geological data © Geological Survey of Finland.

Palaeoproterozoic Central Lapland Area

The Central Lapland area consists of the ca. 2.50 Ga intracratonic rift-related Salla and ca. 2.44 Ga Onkamo Groups (Lehtonen et al. 1998). A depositional hiatus is inferred between these and the first of the transgressional cratonic and/or cratonic margin sequences (>2.21 Ga Sodankylä Group, Lehtonen et al. 1998). The >2.13 Ga to >2.05 Ga Savukoski Group (Lehtonen et al. 1998) sedimentary rocks reflect deepening basin conditions (Lehtonen et al. 1998). Younger rocks of this Group include ca. 2.05 Ga picritic and komatiitic rocks of the Sotkaselkä and Sattasvaara Formations for which a mantle plume origin has been proposed (Hanski et al. 2001). Rocks of the ca. 2.0 Ga (Lehtonen et al. 1998) Kittilä Group are argued to be allochthonous (potentially a forearc sequence), emplaced from the southwest or west over older units as a series of nappes (Hanski 1997).

Central Lapland Granitoid Complex

The Central Lapland Granitoid Complex is composed of migmatitic sediments, volcanic and plutonic rocks. The Complex is characterised by Paleoproterozoic intrusions with a very broad range of ages. Archean dates in the southeast of the Central Lapland Granitoid Complex include 2815 Ma and 2754+16 Ma from tonalitic intrusions and ca. 3.4 Ga to 2727+13 Ma for detrital zircons in adjacent quartzites (Evins et al. 1997, 2000). Inconclusive Archaean ages have been found for dacitic and rhyolitic volcanic rocks near the northern and eastern margin of the Central Lapland Granitoid Complex (2721+9 Ma, 2746+14 Ma Rastas et al.2001; 2775+25 Ma Räsänen & Huhma 2001; 2800+8 Ma Räsänen & Vaasjoki 2001).

Near the northern border of the Central Lapland Granitoid Complex, 2.1 Ga granites (Huhma 1986) intrude an existing deformation fabric and regional metamorphic mineral assemblage. Younger intrusions belonging to pre-, syn-, and post-orogenic groups include the ca. 1.89 Ga to 1.86 Ga syn-orogenic intrusions; 1.84 Ga late orogenic granites and gabbros; and the ca. 1.80 Ga Lohiniva Suite and post-orogenic mafic to monzonitic/syenitic rocks and microcline granites/appinites (Väänänen 2004). Post-orogenic granitoids of age ca. 1.80 Ga to 1.77 Ga have also been described.

Peräpohja Schist Belt

Two main litho-stratigraphic groups (lower Kivalo and upper Paakkola Groups) have been defined for the Peräpohja Schist Belt. The basal conglomerates of the Sompujärvi Formation unconformably rest on both the Archaean Pudasjärvi Complex and the 2.44 Ga layered intrusions (Perttunen & Vaasjoki 2001). The continental-type mafic metalavas of the Runkaus Formation (bracketed between 2.44 Ga and 2.22 Ga) above the Sompujärvi Formation show evidence for contamination by Archaean crust. Quartzites of the overlying Palokivalo Formation constitute the greatest volume of preserved rock within the Peräpohja Schist Belt, and are composed of primarily of Archaean-sourced material. The voluminous oceanic-type basalts of the subsequent Jouttiaapa Formation (c. 2.1 Ga) indicate advanced rifting, as these show no geochemical indications of interaction with Archean crust or Archaean sub-continental lithosphere. However, the basalts of this Formation are overlain by cratonic Kvartsimaa Formation orthoquartzites and dolomites. The Paakkola Group containts politic black schists and minor mafic volcanic rocks. Additional formations have been defined for the northern transition zone between the Peräpohja Schist Belt and Central Lapland Granitoid Complex. These include the Korkiavaara Formation, containing interlayered 'arkosic' and mafic volcanic rock layers. Several mafic dyke generations occur along the southern margin of the Peräpohja Schist Belt, with broad age groups of 2.44 Ga, 2.20 Ga, 2.10 Ga, and 1.98 Ga. The minimum age of rocks in the Peräpohja Schist Belt is defined by the 1.88 Ga age of Haaparanta Suite intrusions that intrude all of the Peräpohja Schist Belt rock types. Metamorphic grade increases from greenschist to amphibolite facies northward towards the Central Lapland Granitoid Complex.

Kuusamo Schist Belt

According to Silvennoinen (1991), the stratigraphic sequence of the Kuusamo Schist Belt started with the deposition of a thin, discontinuous polymictic conglomerate unit directly on Archaean crust. This was followed by Greenstone Formation I, which chiefly comprises mafic and intermediate amygdaloidal and massive lava units with tuffitic interlayers. The overlying Sericite Quartzite Formation is composed of sericitic quartzite, feldspathic quartz-pebble conglomerates, sericite schists, carbonate rocks (dolomitic marbles) and felsic volcaniclastic rocks. The extensive Greenstone Formation II basalts cover the Sericite Quartzite formation. This of overlain by the Siltstone Formation, which now is composed of meta-arkose, orthoquartzite, phyllite and dolomitic marble members. This was followed continental tholeiitic basalts of the Greenstone Formation III, and the Rukatunturi Quartzite Formation comprising sericitic and arkosic quartzites and orthoquartzite. These, in turn, were covered by the Dolomite Formation which separates the quartzites from the uppermost stratigraphic unit of the Kuusamo Schist Belt, the volcanogenic Amphibole Schist Formation. Differentiated dolerite sills and dykes, 'albite diabases', cut all formations except the Amphibole Schist Formation. These mafic intrusives belong to two age groups, 2206±9 Ma and 2078±8 Ma (Silvennoinen 1991).

Archaean Pudasjärvi Complex

The Archaean Pudasjärvi Complex comprises mainly complexly deformed gneisses and remnants of greenstone belts. Furthermore, a belt of layered mafic intrusions (2.44 Ga) straddles along the margin between Archaean and Proterozoic Peräpohja Schist Belt. The Archaean Pudasjärvi Complex is dominated by Archaean felsic paragneisses and orthogneisses, including tonalite-trondhjemite gneisses and migmatites. Sub-elements including Archaean greenstone and granulite belts have been identified in the southern parts of the Complex (e.g., Oijärvi Greenstone Belt, Perttunen & Vaasjoki 2001, and Pudasjärvi Granulite Belt, Mutanen & Huhma 2003, Huhma et al. 2004).

Pudasjärvi Complex rock ages vary from ca. 3.5 Ga to 2.7 Ga (as shown by the age of the Siurua gneiss within the Pudasjärvi Granulite Belt and of the Ranua diorite respectively, Mutanen & Huhma 2003). The Siurua gneisses are currently the oldest known rocks within the Fennoscandian Shield, and an even older inherited zircon core age of 3.73 Ga also exists for these rocks (Mutanen & Huhma 2003). Age data for the Pudasjärvi Granulite Belt suggest several Archaean crustal formation stages from ca. 3.5 Ga to 2.8 Ga (Huhma et al. 2004).

Mineralization

Some nickel-copper sulphides and VMS-type showing has been discovered thus far in the Archaean greenstone belts, whereas orogenic gold occurrences are more abundant and are part of an important global mineralising event at 2.7 Ga. The fragmented 2.5-2.4 Ga layered mafic-ultramafic intrusive complexes within the Karelian craton host major chromitite, nickel-copper and platinum group metals (PGM) deposit. The Archaean crust underwent numerous rifting events at 2.4-2.0 Ga, leading to voluminous ultramafic to mafic volcanism and plutonism in central Finnish Lapland. Resulting nickel-copper mineralizations occur in host rocks that vary in composition from komatiite-type to ferropicritic.

In northern Finland, dozens of orogenic gold occurrences have been discovered in the extensive Palaeoproterozoic greenstone belts of Central Lapland, Kuusamo and Peräpohja. Some of the orogenic gold deposits in northern Finland have an atypical metal association although most of their features are similar to the gold-only orogenic occurrences. They are significantly enriched in copper and, in some cases, also in cobalt, lanthanides/rare-earth elements, nickel and/or uranium, and are surrounded by intense albitization predating gold mineralization and related alteration. There are also iron oxide-copper-gold (IOCG) deposits in northern Finland. The most significant occurrences are the Kolari magnetite ore bodies in westernmost Finnish Lapland at the Swedish border.

Pre-Quaternary Regolith

Weathered bedrock surface has been preserved beneath glacial deposits in many areas in northern Finland (Hirvas 1991). Up to tens of meters thick remnants of weathered bedrock are frequently found particularly in topographic depressions under till in Central Lapland. Typically only the saprock has been preserved, but in places also the lower saprolite and parts of the upper saprolite of the weathering profile are present (Sarala et al. 2007a). The saprock horizons are strongly fractured and they have been zones of preferential groundwater movement and they are commonly rich in secondary iron minerals like goethite. Trace elements such as Cu, Ni, Co, Zn and Mo have been enriched in the fine fraction of the goethitic weathering crust and the concentrations can be many times higher than in the underlying fresh bedrock (Peuraniemi 1990a). The mixing of the weathered material in till causes problems in till geochemistry and can in places distort the geochemical pattern. Large amounts of secondary enriched weathered material in till may not necessarily be related in any way to ore deposits.

In the areas where active ice lobes have been existed, the remnants of weathered bedrock are mostly eroded away. For example in southern Lapland Pre-Quaternary regolith is rare. The same situation is also in the northernmost parts of the Lapland. This suggests that the glacial activity in the central Lapland area and the other areas on the northern and southern side of it were different. It also indicates that the central Lapland area repeatedly existed in the central part of continental glaciers. In this ice-divide zone, the cold-based conditions were dominant and it was poorly eroding during the late deglaciation phases as pointed out by Hirvas (1991).

Based on the illite dating results presented by Sarapää (1996), weathering reached the current erosion level about 1,200-1,000 Ma ago in the southern Finland and the weathering profile probably developed episodically till Neogene. This is much earlier than the ca. 100 Ma suggested by Hirvas and Tynni (1976), Söderman (1985), Saarnisto and Tamminen (1987).

Surficial Geology i.e. Quaternary Glacial Geology

Finnish Lapland has located repeatedly in the centre of Fennoscandian continental glaciers during the Quaternary period. The core areas of glaciers existed mainly in the Norwegian and Swedish mountain areas but also in northern Finland. In the eastern sector, the ice margin proceeded as far as the northwest Russian Plain during the Late Weichselian Maximum (LGM; 25,000-9,000 years ago). During the Early (115,000-75,000 years ago) and Middle (75,000-25,000 years ago) Weichselian several glacial phases existed but not in such a large volume (cf. Svendsen et al. 2004). For example Mäkinen (2005), Sarala (2005a), Salonen et al. (in press), in their recent papers, presented that ice-sheet was covered northern and initially also central Finland during the Middle Weichselian but not earlier. Instead, northern Russia and North America were largely covered by ice sheets during the Early Weichselian.

Hirvas (1991) presented three different till beds of the Weichselian age in northern Finland. They are related to three different glacial advance phases of which the oldest presents the main glacial phases and the upper one oscillation phases during deglaciation.

There are also observations of three older till beds that represent glaciations older than Weichselian. For example, in the open pit of old Rautuvaara iron mine, in Kolari, Hirvas (1991) was described and interpreted six different till beds in the same section. Ice-flow directions are different in the different till beds, which is remarkable thing for the mineral exploration and should be considered in the planning of the till sampling.

On deglaciation phase strong warming and rapid climate changes were the reasons for quick melting of the ice sheet both on the marginal part and on the surface of ice, which led to the changes in glacial dynamics, ice lobation and ice-flow directions. In this environment rapid changes between erosional and depositional conditions have led to variable degree of preservation of earlier deposits and pre-Quaternary regolith, and deposition of complex glacigenic deposits. Last ice divide zone was in the Central Lapland (Figure 2) where the ice-sheet separated and begun to flow to the northeast-wards and also in the southeast-wards. Two large drumlin fields in the Inari region (ice-flow from the southwest) in the northernmost Finland, and in the Kuusamo area (ice-flow from the northwest) are clear examples of the ice sheet separation and fast flowing ice streams. This indicates that precipitation and followed accumulation in the central area of glacier were increased causing high internal pressure into glacier, and followed fast flowing ice-streams (Sarala 2005a).

Figure 2: Glacial landform provenances of northern Finland and the location of last ice-divide zone in central Lapland area. Main ice-flow direction during last deglaciation is indicated with arrows. White area is mainly composed of till and other glacigenic deposits. The main target areas and other places mentioned in text are also marked on the map.

Northern Finland became ice free about 11,600-10,000 years ago (Saarnisto 2005). Ice margin was ended into time-transgressively developed ice-dammed lakes or large Ancylus Lake that bordered ice margin in southern Lapland area. That Ancylus Lake phase continued also during postglacial time and changed via short Littorina Sea to the modern Baltic Sea phase.

METHODS IN SURFICIAL GEOLOGICAL EXPLORATION

An initial method in surficial geological exploration is an identification of glaciogenic landforms (Salonen 1988, Aario & Peuraniemi 1992, Sarala 2005a). The use of aerial photos and satellite images is a traditional way on interpreting glacial morphology. Nowadays the digital elevation model (DEM) is also practicable as a geomorphological interpretation tool due to flexible use in different scales and on the GIS analysis. After that, glacial morphological maps can be composed. In this phase, rough estimations of glacial transport distances and directions can be done.

On the second phase, the characteristics of different moraines and landforms, e.g. occurrence, relation to the other forms, composition, inner structure and till stratigraphy must be studied (Hirvas & Nenonen 1990). This needs usually field work with test pit excavations, natural or man-made-section observations or drillings with sampling. Till geochemical and heavy mineral samplings are essential in mineral exploration but also sampling for grain size determinations and pebble countings. Fabric analysis and striae observations are still necessary for the determination of ice-flow direction and in till stratigraphy. Fabric analyses can also applied in determination of an amount of primary and secondary components that indicate depositional environment of till unit, if the dip of unique stones is also measured in addition to the direction. By this way, the range and variation of depositional processes of glacial landforms both in horizontal and vertical dimensions can be done (Sarala 2005b).

Finally, the glacial dispersion, transport distances and sedimentation processes as well as the history and glacial dynamics of the last glaciation can be estimated by combining the various morphological data (e.g. erosion marks, glacial landforms, and glaciofluvial and -lacustric deposits), stratigraphical data (described earlier in this chapter) with complementary bedrock studies and geophysical surveys (Sarala et al. 2007a). A multidisplinary approach using various exploration methods allows better interpretation and analysis of the results.

Till geochemistry is one of the key methods for the surficial mineral exploration. Till sampling can be carried out by many different ways depending on the scale and purpose. Regular sampling grids are used in the regional and local scale approach, when the idea is to map the variation of the elements in the study area. This sampling is easiest to do by percussion drilling with a flow-through bit or by other light weight methods (e.g. by spade or hand drill). Line sampling is useful, when the effect of regional soil and till variations has to be minimized, or when the bedrock structure and ice-flow direction are known. Test pits and test trenches are as their best in detailed studies, when the sources for anomalies are needed to clarify. In other hand, the test pits excavations should be the first field survey in a new study area to find out the till stratigraphy and the thickness of glacial drift in the area (Sarala 2005a). This knowledge is useful later when planning sampling strategies or other exploration studies like diamond drilling or erratic boulder mappings. Vertical variation of the elements can be studied by taking samples at regular intervals from different depth levels. If test trenches are dug, both horizontal and vertical behaviour of the elements in glacial drift can be determined using sampling profiles at regular distances.

Another useful method in exploration is a heavy mineral study (Peuraniemi 1982, 1990b, Lehtonen 2005, McGlenaghan 2005), which is based on the separation of the heaviest mineral fraction from the till samples. Sample size can vary depending on the purpose of the study but usually ten or twelve litres is enough to get a representative sample of the heavy mineral composition. Rough concentration of the samples can be done with a spiral separator (gold screw panner) or shaking table, or by panning. After removing the magnetic fraction, heavy liquids (density 2.96 g/cm^3 or 3.31 g/cm^3) can be used to separate the heavy minerals. Basic heavy mineral examination is done using a binocular microscope or an electron microscope (EDS+SEM) and detailed mineralogical study using a microprobe (EMP). Heavy mineral concentrates are useful especially for metallic gold determination but also for other metals, heavy silicates, garnets, chromite, zircon and sulphide minerals.

GLACIGENIC DEPOSITION

Glacial dynamics and subglacial conditions are the main factors affecting on erosion, transportation and deposition. An identification of prevailed sub-glacial conditions, i.e cold-based, warm-based and partly frozen/thawed-bed conditions, during the deposition is essential task. In glacial dynamics' point of view, on cold-based conditions movement occurs only as an internal deformation while basal movement does not happen (Eyles & Menzies 1983, Eyles & Eyles 1992, Kleman 1994). In cold-based conditions older landforms can preserved (Kleman et al. 1994, Clark 1999). When glacier's mass and volume increase, and basal conditions change to warm-based, glacier begins to move by basal sliding. This is the phase where older overburden erodes away, and when the abrasion begins. Warm-based conditions are common on outer marginal part of the glacier (Aario 1990, Kleman 1994). In the inner marginal part warm-based conditions change to cold-based and basal conditions are partly thawed and frozen. In this zone basal ploughing is possible and large erratic boulders can be quarried from the bedrock surface (Sarala 2005b). Large block movements are also possible under fast streaming ice-lobes. In this case, the ice is staged on the ground and the fast-flowing ice, when the movement reach the ice bottom, creeps the ice-sediment blocks moving them some tens to hundreds meters. Those sub-glacial conditions have direct influence on till stratigraphy and glacial morphology.

Northern Hemisphere has been covered numerous times by continental ice-sheets. The conditions have been like in Antarctica or Greenland nowadays. It is worth remembering that those conditions both under ice and on the margin have differed totally of those that exist in modern glaciers, valley glaciers or on the mountain tops where most of the investigations for glacial dynamics have been carried out. Ice masses of size of continental ice-sheets have been huge and glaciers' behaviour and dynamics have been independent from the many outside factors like underlying topography and annual climate changes. Latest investigations have also proved that the growth of ice masses to the size of continental glaciers have been rapid event, and have lasted only 5 to 7 thousands of years as a minimum for example in Fennoscandia and northern Europe (Saarnisto & Lunkka 2004, Svendsen et al. 2004). And, as the growth was rapid, also the melting phase was very short. For example the melting phase of the Weichselian glaciation lasted only 7 to 8 thousands of years and was affected by enormous changes in climate (Bard 2002, Sarala 2005a). That is one reason to the division of ice margin into active and passive ice-lobes.

Tills and Glacial Landforms as Indicators of Transportation

There are two different ways to analyse prevailed depositional conditions: 1) till facies analysis or 2) glacial morphological analysis. Identification of different till units is based on the analysis of composition and structure of till deposits. The key point is to recognize basal tills like basal lodgement and melt-out tills of those that are indicating deposition on marginal part of glacier. Typical tills that indicate marginal deposition are flow till and deformation till, which have affected by secondary, usually non-glacial processes (e.g. gravity flow) during the deposition. Dreimanis (1989, 1990) has described characteristic features of different till types and classification of till deposition under influence of primary and secondary processes. Dreimanis is also presented several other special till types that indicate variable deposition conditions in divergent glacial environment, and what should be keep in mind when interpreting the results.

Another way is to interpret the glacial morphology and apply that knowledge for recognizance of depositional environment. In the case of exploration it is important to distinguish active-ice deposits from passive-ice ones. Active-ice deposits are those that are formed sub-glacially under advancing ice sheet, like drumlins, flutings, ribbed moraines and thin basal moraine sheet. Instead, hummocky moraines and end or other marginal moraines, and thick ablation moraine formations are deposited under passive-ice conditions and usually during the retreating phase of glacier. The main factor that separates environments from each others is the transportation mechanisms of source material for till. Till types are usually lodgement or basal melt-out tills in the case of active-ice deposits, so the tills of those landforms are deposited sub-glacially, by grain by grain or fragment by fragment, under high pressure. In spite of great variation between the transport distances in different formations, the transport direction is clear and in the line with ice-flow direction. Till in passive-ice formations are usually transported en-glacially or even supra-glacially, so the transport distance and direction can vary a lot and the source area(s) are not easily traced. The same features are typical for glaciofluvial and -lacustrine deposits, which are not handled in this article.

Streamlined, Active-Ice Landforms in Indicator Tracing

Drumlins and flutings, and some other radial moraine types are very clearly indicative of ice-flow direction. These landforms form typically large streamlined moraine fields where the number of single forms can be several thousands. The dimensions of drumlins are usually 5-50 m in height, 10-200 m in width and from 100 m to several kilometres in length, and they are mainly formed under warm-based glacial conditions on marginal zone of ice sheet. These forms can be both depositional and erosional. Depositional forms are consequence of anisotropic differences in the sub-glacial debris when plastic debris is mobilised due to chancing stress fields or the influence of active basal melt-water which carves cavities beneath an ice mass that are later infilled with stratified sediments and till (Menzies & Shilts

1996). Moulding of previously deposited material in sub-glacial conditions, where the meltwater content is low, is an explanation for erosional forms. For that reason, in this case, the sediment bedding and structures usually represent the earlier depositional history of the area.

Flutings are mainly formed in a similar way as depositional drumlins but their dimensions are much smaller, height is 1-2 m, width 2-10 m and length 100-1 000 m (Aario 1990). These forms are usually easiest to notice on aerial photos as a group of narrow stripes in close connection with drumlins. Identification of the forms is difficult in the field if vegetation covers them. Furthermore, in Finland they are usually surrounded by the peat bogs, of which surfaces are almost on the same level with flutings tops.

The transportation of till in streamlined features is much investigated in the course of years and several symposiums were organized including this subject (DiLabio & Coker 1989, Aario & Heikkinen 1992, Piotrowski 1997). Transport direction of mineralized till debris and boulders is usually converges with the orientation of drumlins and flutings but transport distances are quite long: usually several kilometres, in places even tens of kilometres. Aario and Peuraniemi (1992) have classified the applicability of these forms usually good to exploration. That is case when only one clear advance phase occurs, and ice movement is simple and straightforward on the ground, where older glacial overburden is relatively thin. But if multiple glacial advances with variable flow directions exist, glacial dispersion can be very complex and transport distance long. For example in Kuusamo drumlin field, in northeastern Finland at least two different flow phases have been observed (Aario & Forsström 1979). Also, in the Central Lapland area, on northern side of the last ice-divide zone where ice has flown towards the northeast during the last deglaciation, Sutinen (1992) has described drumlins from older Weichselian glacial advance phase with totally different northwest-southeast orientation. This is possible because the location of ice-divide zone of earlier Weichselian glaciation has been about 100 km to the north from the latest ice-divide zones (c.f. Hirvas 1991). Changes in the location of glacier's centre and ice-divide zones are possible during the different glaciations and even during the different glacial phases of one glaciation, and should be considered when interpreting till geochemical results, particularly on areas of complex till stratigraphy.

Latest studies in Korkea-aho at the Kuusamo drumlin field, support opinion of the long transport distance of mineralized material in drumlins. The bedrock in this area is composed of Archaean sedimentary and volcanic rocks of Kuusamo Greenstone Belt. In four test pits dug on the proximal part of drumlin gold contents were elevated (Figure 3). Drumlin (width 1 km, width 200-300 m, height 15 m) in Korkea-aho is composed of two till units, of which the upper one is distinguished as brownish-grey, sandy, matrix-supported till having generally compact and massive structure. Some fissility structures and thin sandy or silty stripes are a mark of sub-glacial deposition. The rock composition of the roundness pebbles indicates long glacial transportation. The lower till unit is grey and has silty or sandy matrix. The till is slightly laminated in places but otherwise has mostly massive structure. The lithology and roundness of pebbles is a same than in the upper till. Fabric countings were made only from the upper till unit showing the same orientation with drumlin body in the topmost part of the ridge (Figure 3). At marginal parts of the ridge, fabrics showed that the stones have turned gently away from the drumlin's centre line (about 290°). It indicates that pressure during the deposition has been from the drumlin's body towards the depressions on both sides of the drumlin, and that the drumlin is depositional landform. The nearest known gold mineralization exits on the western and north-western side of the Kuontijärvi Lake, which means that transport distance is about five to eight kilometres. A source can also be under the lake but the transport distance is still several kilometres.

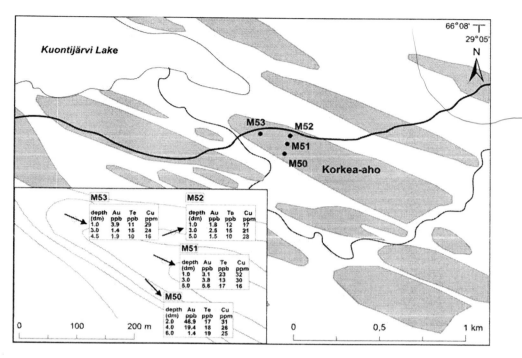

Figure 3: Topographic map of the drumlins (grey) in Korkea-aho, Kuusamo, north-eastern Finland. Till geochemistry (Au, Te and Cu in fine size fraction <0.06 mm) and till fabric of the upper till unit in different test pits (M50-M53) are presented on lower left corner.

Transverse, Active-Ice Moraine Ridges in Indicator Tracing

Only transverse, active-ice moraine type in northern Finland is ribbed moraine. The ribbed moraine morphology consists of transversal (relative to the ice flow direction) till ridges, which dimensions are 100-1 000 m in length, 50-200 m in width and 2-10 m in height. The interval of individual ridges is 100-300 m. The term ribbed moraine is sometimes used as a synonym for Rogen moraine but here it is used as the name for all moraine ridge types formed by a similar process and with a similar morphology (cf. Sarala 2003).

Hättestrand (1997) has divided ribbed moraines into four subtypes: hummocky ribbed moraine, Rogen moraine, Blattnick moraine and minor ribbed moraine. All of the types have been formed as a result of quite similar conditions and formation processes during the deglaciation, under a fast flowing glacier at the contact between cold-bed and thawed-bed. Hättestrand (1997) and Kleman and Hättestrand (1999) suggest that due to the tension subglacial deposits broke up and moved with the flowing ice. As a result the undulating, puzzle-like ridge morphology was formed. Sarala (2005a) suggests that the ribbed moraine formation was a two-step process: 1) initially a strong tensional pressure that was caused by strong internal ice movement fractured the cold-bed glacier and the sub-glacial sediments into blocks which moved under the ice sheet and formed ridge morphology, and then 2) the flowing ice transported the material loosened between the blocks by the freeze-thaw process and redeposited it on the surfaces of the new ridges. The deposition has occurred in the transitional zone between the cold-bed and the thawed-bed glacier during deglaciation. By this mechanism, the surface boulders typical of ribbed moraines and the abundant occurrence of local boulders in the uppermost till unit can be explained.

Ribbed moraines have been used in exploration in southern Finnish Lapland area and a lot of results have been published recently (e.g. Sarala et al. 1998, Sarala & Rossi 2000, Sarala 2005a, Sarala & Peuraniemi 2007). The excellence of this moraine type for exploration is that the composition of debris and rock fragments in upper part of the ridges, and the surficial boulders on the ridge top indicate strongly the composition of underlying bedrock. It makes exploration easy and fast because the surficial boulder tracing is straightforward task and till geochemical anomalies are easily detectable due to narrow and sharp secondary dispersion. This is clearly seen for example in the case of Petäjävaara Cu-Au mineralization in south-western Rovaniemi (Sarala & Rossi 2000), which situates in the Paleoproterozoic Peräpohja Schist Belt. The Cu-Au mineralization was found in hydrothermal alteration zone in between quartzite and mafic volcanic rocks, and is related to one-metre wide quartz vein with associated pyrite and chalcopyrite dissemination. The transport distance of mineralized boulders and pebbles, and also for a main part of mineralized till debris is very short (only 5-50 m; Figure 4). Similar exploration cases with very short glacial transportation have been reported also in the Portimojärvi area, north from Ranua village (Aario 1990, Sarala & Peuraniemi 2007), and in Misi area, north-eastern Rovaniemi (Sarala & Nenonen 2005).

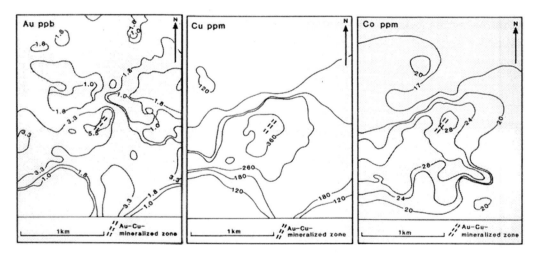

Figure 4: Distribution of Au, Cu and Co in fine till size fraction (<0.06 mm) in ribbed moraine area in the Petäjävaara target area, in southern Finnish Lapland as an example of short glacial transportation (Reprinted from Sarala et al. (1998) with permission of Geological Society of Finland).

Till Stratigraphy

Thickness and the number of till units and till beds vary in relation to the sub-glacial activity, which is, as mentioned earlier, depending on the position of deposition environment under ice-sheet in between the ice-edge and centre point. In the central area, like in places in central Lapland, where glacial erosion has been weak, glacial overburden is usually thicker than in the active marginal parts and fast-flowing ice-lobe areas. Also, the stratigraphical record represents usually longer history than in the shallow overburden areas. Glacial overburden is composed of several till units, which represent different glacial advance phases with variable stone orientations and rock compositions. Unfortunately, at the same time it also means that till stratigraphy is complicated (cf. Hirvas & Nenonen 1990).

Till stratigraphy is usually quite simple in the case of active ice-lobes. Older deposits can be totally eroded away and deposited as new till units representing advance and retreating phases, which as together form one till bed. In that case, the lower till is usually indicative of underlying bedrock composition and for example, till geochemistry and heavy mineral composition is reflecting the element variation in relation to glacial transport distance. The upper till, instead, is somewhat largely reflecting element variation in the bedrock. This is due to en- and supra-glacial origin of rock fragments, debris and heavy minerals in till that are largely dispersed and transported.

However, glacial overburden is rather as simple as described above. Because of long glacial history and variations in ice-flow directions, tills from other glaciations or other glacial advance phases remain in topographic depression and on distal end of topographic obstacles. Furthermore, glacial erosion can sometimes favoured ploughing instead of abrasion of which the redeposited till or other sediment blocks are remarks in the latest till units. Glacier can also move large sediment plates along the ice bottom and redeposit them into wrong stratigraphical horizon in overburden. It is also good to notice that till is, at least in the repeatedly glaciated areas, always a mixture of reworked glacigenic or other, older sediments, and fresh bedrock material. The proportion of each component can be determined by using wide range of sampling and by studying physical parameters of till fragments.

The deposition process of till can be in some cases opposite related to simple stratigraphy example presented earlier. In the ribbed moraine case the local bedrock composition is seen in the uppermost till unit. This is due to very effective quarrying during the sub-glacial deposition under freeze-thaw process (Sarala 2005b). Same type of phenomenon can occur on the hilly environment when ice erodes bedrock surface on the hill slopes and deposits that fresh material on the lower topographic level, above the earlier deposited till sheet.

The relation of till geochemistry and till stratigraphy is only possible to study if samples are taken frequently from each till units. For example, half metre interval is enough to recognize differences in between the till units. The difference between the till geochemistry of fine size fraction (< 0.06 mm) of upper and lower tills is great in ribbed moraines and drumlins that are deposited very different ways (Figure 5). Also, both the horizontal and vertical variation is possible to examine if the sampling tension in both dimensions are studies and the sampling grid is tense enough (see the example from the Vammavaara target area in Sarala 2005b).

An effect of till stratigraphy is tested by using two different sampling methods in our studies. Test pit samples are taken under high control, and for that reason, the relation into the stratigraphical units can be pointed out. In the case of percussion drilling, it is hard to say exactly which till bed the samples are taken from (colour, grain size, lithology etc. can be the same). Difference between bedrock and big erratics or stone pavements in till is also difficult to notice, particularly, if only one sample is taken from as deep as possible (cf. Peuraniemi 1982).

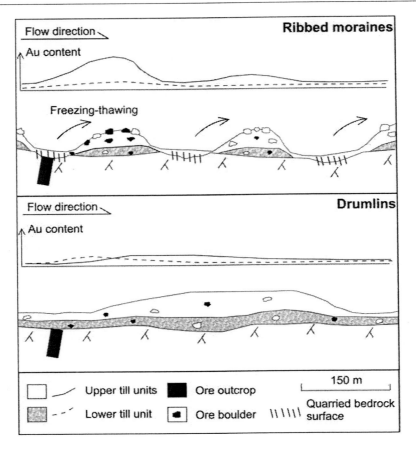

Figure 5: Difference between the transport mechanisms and distances of mineralized material in till in ribbed moraine ridges and drumlins in southern Finnish Lapland (Reprinted from Sarala & Nenonen (2005) with permission of Geological Survey of Finland).

TILL GEOCHEMISTRY AND HEAVY MINERALS IN EXPLORATION

As pointed out in the beginning of this article, till geochemical and heavy mineral exploration are effective methods in glaciated areas. Usually, mineralizations if occurring as outcrops in the bedrock surface cause some kind of ore metal/element anomalies, or at least an indicator element anomaly into the till beds. In tracing the source, the sampling density is an important factor as it is discussed broadly in the literature (e.g. Gleeson et al. 1989, Koivisto 1992, Hartikainen & Nurmi 1993). The levels of sampling density from regional to detailed are depending on the stage of exploration studies.

Till geochemical investigation is often quite cheap and effective method in targeting the exploration. Till samples can be collected by mobile sampling equipment from the surface of glacigenic overburden, and after that the samples are ready for analysis. In mobile metal ion (MMI) method the samples can be taken even by a small spade or hand drill. Heavy mineral study, instead, is somewhat complicated method, because the sampling in till cover is only possible if the test pits are dug. The benefit is, however, that the stratigraphical accuracy is higher in the case of heavy mineral samples. Till geochemical samples are, of course, sensible to take at the same time. Sometimes heavy mineral study can be used as a quick overview method for looking gold potential of the target area by using till or weathered bedrock samples that are concentrated by for example light-weight gold screw panner or table shaker in the field, and then examined visually or by using microscope. This is suitable method if only gold nuggets are needed to be count. Usually heavy mineral concentrates are studied more detailed, which requires many pre-processing steps before the binocular microscopic or electron microscopic determinations are possible.

Dimensions of Till Geochemistry

Secondary glacial dispersion can be examined in different scales. DiLabio (1990) has used continental, regional, local and property-scale classification, where the transport distance varies from over 100 km to under 100 m. On the other hand, it means that till geochemistry is only useful on local and property scales, sometimes perhaps on regional scale (cf. Salminen 1995), because the elements concentrations dilutes during long transportation and small amounts are hard to separate from the background. Only erratic boulders are possible to distinguish far away from the source areas.

The shape of dispersal train can be complex in the case of multiple glacial advance phases, i.e. at the case of complex till stratigraphy. Hirvas and Nenonen (1990) and Klassen (2001) have discussed of the shape of trains, and like the glacial morphology, the shape can also be useful when reconstructing glacial mechanisms and environment during the deposition. Furthermore, dispersal trains are also vertically controlled. DiLabio (1990) presented that dispersal trains rise gently towards the down-ice direction. The anomaly is coming out at the surface after 300-1000 m transportation in shallow till area. This is case when the basal glacial conditions are quite stabile and till is deposited under the lodgement process. In the central part of continental glacier, on variable sub-glacial conditions, transport distances can vary greatly. For example in the drumlin areas in northern Finland till anomalies are narrow but long and mineralized boulders can be found far from the source(s) (usually from several kilometres to tens of kilometres; Aario & Peuraniemi 1992). Contrary to that till anomalies are distinct and short at the ribbed moraine areas in southern Finnish Lapland (Sarala et al. 1998, Sarala & Rossi 2000, Sarala 2005a, Sarala & Peuraniemi 2007). For example in the Petäjävaara target area, described earlier in *Glacigenic deposition* chapter, mineralized material (Cu-Au-bearing boulders and till debris) is seen anomalous not farer than five to ten metres on distal side of mineralization in the bedrock (Figure 6).

In the ice-divide zone the glacial erosion and transportation is usually weak. But due to division of the glacier margin into ice-lobes also during the latest melting phases, active lobes can occur in the ice-divide zone too. That has been the case for example in the Petäjäselkä target area, in central Lapland area, where gold exploration is on-going (Sarala et al. 2007b, Nykänen et al. 2007). In the Petäjäselkä target area the gold anomalous zone follows a heterogeneous and deformed NNW trending zone of graphic tuffs, cherts and intermediate volcanic rocks within a mafic volcanic rock dominated domain. Till stratigraphy is simple: above fresh bedrock is one to two metres thick weathering crust which is composed of fractured bedrock surface or deeply weathered saprock crust, it was followed one to three

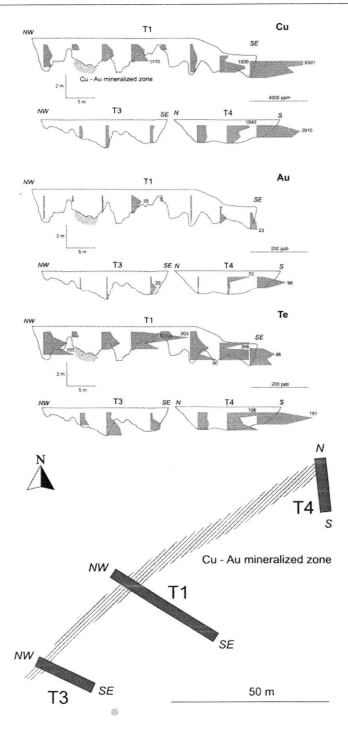

Figure 6: An example of short glacial transport of mineralized material (Cu, Au and Te in fine till size fraction (<0.06 mm)) during the deposition of ribbed moraine ridges in Petäjävaara target area, southern Finnish Lapland. Modified after Johansson and Nenonen (1992; unpublished report in GTK's database).

metres thick till bed that includes two till units; the bottommost till deposited during the advance phase and the upper unit deposited as ablation till on the melting phase. Till geochemistry shows very rapid upward dispersion of mineralized material from the bedrock surface to upper till after 15 m glacial transport from the west to the east (Figure 7). The anomaly pattern is very sharp and strong, and indicates nicely gold mineralization in the underlying bedrock. In recent diamond drillings over 25 g/ton contents have been analyzed from the best part of hydrothermal alteration zone in the bedrock.

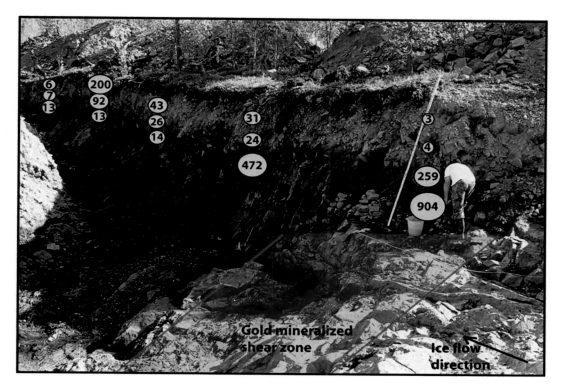

Figure 7: Gold contents (ppb) in <0.06 mm size fraction in till on the test trench M9 in Petäjäselkä, central Lapland, showing rapid upward dispersion of mineralized material from the Au mineralization to the upper till. Photo P. Sarala.

Dilution of Mineralized Material in Till

Glacial dispersion is not only parameter in till geochemistry that can be used in transport distance estimation. The dilution of element contents, i.e. falling of the element contents in a level of background contents (Larson & Mooers 2004), tells also a lot of glacial transportation and transport distances. Dilution of element contents has straight connection to the comminution rate of mineralized fragments in till. This means that fresh crystallized rock fragments last much longer than weathered and already sheared or otherwise softened fragments. For example in the case of test trench at the Petäjäselkä target area gold contents in the upper till sequence on the western part show background values (3-4 ppb) typical for the area (Figure 7). However, after twenty metres transport distance, gold contents are already as same level as background contents and in the profile point 30 m from the mineralization (not seen in the photo) once again in the background level (4 ppb).

Strong dilution is particularly seen in till geochemistry of ribbed moraines. Rapid falling of the contents into background level is almost as quick event as the elevation of content after mineralization. Sarala and Rossi (2000) and Sarala and Peuraniemi (2007) have pointed that the dilute distance is only some tens of meters to 100 metres for gold and a few hundreds of metres for copper and other indicator elements (Figure 4). In the case of ribbed moraines the rapid dilution is seen also in heavy minerals, pebbles and erratic boulders with same length of transport than in till debris. Aario and Peuraniemi (1992) for example have given a very good example of dilution and glacial dispersion in uranium exploration studies in Kuohunki, northern part of the Ranua ribbed moraine field, southern Finnish Lapland.

Figure 8: The contents of Au, Co, Cu, Fe, Bi, Sb and Te in fine till size fraction (< 0.06mm) at the test pit 2006 POS-000098 at the Petäjälehto target area, northern Finland. Topmost row (*) shows background concentrations of the elements in till fines on the Archaean granite gneiss terrain and (**) in soil in Finland (Koljonen 1992). Photo P.Sarala.

Indicator elements can also be suitable for determination of dilution and secondary dispersion. Dispersion pattern is often larger for indicators than the main element(s) because of the halo effect in the bedrock, i.e. indicator elements have spread larger area than main ore element in the bedrock. It is usually benefit because the indicator element dispersion is easily detectable in till geochemical sampling. An example of this is the latest studies in northern Finland, at the Petäjälehto target area, where the anomalous gold contents in till were traced in Archaean granite gneiss area cut by main southwest-northeast and southeast-northwest oriented faults and shear zones. The bedrock was also cut by the magnetite-bearing shear zones in north-south direction. According to Pulkkinen et al. (2007) these magnetite rich zones in connection with southeast-northwest oriented faults are the most potential for gold mineralization. Till geochemical studies prove that gold is in places anomalous (ranging from 4 to 50 ppb) in till fines but the concentrations are not high in every places. Instead, the pathfinder elements have clear indication of hydrothermally altered and gold-bearing rocks in the bedrock. For example in the test pit 2006 POS-000098 (Figure 8) copper, bismuth, antimony and tellurian have clearly increased contents reflecting gold occurrence in the bedrock. Particularly, bismuth and antimony have huge anomalies on the lowermost samples and are reflecting the source to be nearby. At the next test pits about 50 to 150 m to the down-ice direction, the contents of those elements are still over the background value but diluting quite rapidly as a function of distance.

Heavy Minerals in Exploration

Heavy mineral study is largely used method in exploration because it is effective method in till covered areas. Sample concentrator is easy done in field or laboratory with separators with followed visual or microscopic examination. However, the determination of different type of heavy minerals needs good knowledge of morphology, texture and structure of heavy minerals. Long experience of that work is also benefit.

Averill (2001) has pointed out that the uppermost till bed and the surficial parts of moraine formations are not suitable to heavy mineral studies because the labile minerals like sulphides are weathered from this part of glacial drift. Peuraniemi (1982) presented also that heavy minerals preserve unweathered at their best in the deeper parts, under ground water level. Bloom and Steele (1989) showed that for example Cu-bearing sulphides were decomposed in oxidized samples. So, the samples should be taken from the deeper parts of the glacial overburden if sulphide minerals need to be found. However, in the recent studies in northern Finland sulphide minerals like chalcopyrite and pyrite are also commonly found from the surficial part of till cover. Sarala and Rossi (2000) for example described fresh pyrite grains in the upper till in the ribbed moraine ridges in southern Finnish Lapland. Fresh chalcopyrite grains have also been found near the surface in the same area (Sarala et al. 1998, 2007a, Sarala & Peuraniemi 2007). Fresh pyrite grains with cubic shape and chalcopyrite grains are also found over the ground water level in the central Lapland area (Sarala et al. 2007b, Pulkkinen et al. 2007). These grains are either preserved un-weathered in till or are re-crystallized in suitable conditions above ground water level.

Heavy mineral concentration is commonly composed of silicate minerals, granates, magnetite, ilmenite and hematite, sulphide minerals, zircon, monazite and native gold. Many of them are typical for the hydrothermal altered (Au) mineralization in northern Finland. Gold nugget amounts can be high even in till ranging from 3 to 25 pieces in 12 litres sample. Transport distance is also detectable based on the nugget's morphology, composition and texture. Angular and branched nuggets with rugged surface are indicator of short glacial transport. So are the mixed (quartz-gold) nuggets, too. Instead, rounded and smoothed nuggets indicate long and complex glacial history.

Conclusion

Recent exploration studies have shown that glacial morphology, and erosion, transport and deposition processes must be known for understanding transportation of till material. The knowledge and recognizance of landforms particularly in the central parts of continental glacier is a key to manage exploration. It means that the depositional environment of moraine formations, i.e. active and passive sub-glacial conditions, must be recognized. A rough estimation of transport distances and directions can be done based on identification of landforms and their relations to each other.

The experience from northern Finland shows that the composition and stratigraphy of till cover and the specific characteristics of different moraine formations must be carefully considered before till sampling. Areal test pit excavations with till, regolith and fresh bedrock samplings are the first thing that must be carried out in the new target areas. By this way till sampling can be directed when it covers large area and the sampling grids with frequent intervals have been used. The experiences from the gold exploration in northern Finland show that till samples should be taken from the different depth levels and in the case of detailed investigation by using profile sampling with short intervals in the ice-flow direction.

Glacial dispersion and the dilution of mineralized material in till are useful in estimation of depositional conditions of till and moraine formation. The case studies presented in this article show that deposition processes and transport distance of for example ribbed moraines, drumlins and till cover in the ice-divide zones vary a lot from each other. The determination of transport distance can be supported by heavy mineral studies, fabric and striae analyses, stone countings, grain size analyses, surficial boulder countings, i.e. till stratigraphical analyses.

Till geochemistry is still very useful method in exploration. Dispersion of mineralized material is many times distinct and anomalies are sharp at least in the case of ribbed moraines. In drumlin and ground moraine areas transport distances are usually longer, although exceptions exits in the ice-divide zone where transport distance can be very low (< 50 m). With pathfinder elements the target can be recognized and the target focused for detail studies. In property-scale heavy mineral study is easy and effective method to direct studies, although it needs good knowledge of heavy mineral identification. Studies in northern Finland have also shown that by heavy mineral study sulphide minerals can also be found on the upper till, over the ground water level.

References

Aario, R. (ed.) (1990). Glacial heritage of Northern Finland; an excursion guide. Nordia tiedonantoja, *Sarja A*: 1, 96 p.

Aario, R. & Forsström, L. (1979). Glacial stratigraphy of Koillismaa and North Kainuu, Finland. *Fennia* 157, 1-49.

Aario, R. & Heikkinen, O. (eds.) (1992). Proceedings of Third International Drumlin Symposium, *Geomorphology* 6.

Aario, R. & Peuraniemi, V. (1992). Glacial dispersal of till constituents in morainic landforms of different types. In: Aario, R. & Heikkinen, O. (eds.) Proceedings of Third International Drumlin Symposium, *Geomorphology* 6, 9–25.

Ahtonen et al. submitted.

Averill, S.A. (2001). The application of heavy mineral indicator mineralogy in mineral exploration with emphasis on base metal indicators in glaciated metamorphic and plutonic terrains. In: McGlenaghan, M.B., Bobrowsky, P.T., Hall, G.E.M. & Cook, S.J. (eds.), Drift exploration in glaciated terrain. *Geological Society Special Publication* 185, 69-81. The Geological Society Publishing House, Bath.

Barbey, P., Bernard-Griffiths, J & Convert, J. (1986). The Lapland charnockitic complex: REE geochemistry and petrogenesis. *Lithos* 19:2, 95-111.

Bard E. (2002). Abrupt climate changes over millennial time scales: climate shock. *Physics Today* 55 (12), 32-38.

Barling, J., Marker, M. & Brewer, T. (1997). Calc-alkaline suites in the Lapland-Kola orogen, northern Baltic Shield, geochemical and isotopic constraints on accretion models. In: EUG 9, European Union of Geosciences, 23-27 March 1997, Strasbourg (France). Abstracts of oral and poster presentations. *Terra Nova 9, Abstract supplement 1*, p. 129.

Bibikova, E.V., Mel'nikov, V.F. & Avakyan, K.Kh. (1993). Lapland granulites: petrochemistry, geochemistry and isotopic age. *Petrology*, Moscow, 1, 215-234, (in Russian).

Bloom, L. B. & Steele, K. G. (1989). Gold in till: preliminary results from the Matheson area, Ontario. In: Dilabio R.N.W. & Coker W.B. (Eds.), Drift Prospecting. *Geological Survey of Canada,* Paper 89-20, 61-70.

Clark, C. (1999). Glaciodynamic context of subglacial bedform generation and preservation. *Annals of Glaciology* 28, 23-32.

DiLabio, R.N.W. (1990). Glacial dispersal trains. In: Kujansuu, R., Saarnisto, M. (Eds.), *Glacial indicator tracing.* A. A. Balkema, Rotterdam, 109-122.

DiLabio, R.N.W. & Coker, W.B. (Eds.) (1989). Drift prospecting. *Geological Survey of Canada,* Paper 89-20.

Dreimanis, A. (1989). Tills: their genetic terminology and classification. In: Goldhwaith, R. P. & Matsch, C. L. (eds.), *Genetic classification of glacigenic deposits.* Rotterdam, A.A. Balkema Publishers, 17-83.

Dreimanis, A. (1990). Formation, deposition, and identification of subglacial and supraglacial tills. In: Kujansuu, R. & Saarnisto, M. (Eds.), *Glacial Indicator Tracing.* Balkema, Rotterdam, pp. 35–59.

Evins, P., Ahtonen, N., Airo, M.L., et al. (1997). Preliminary observations of the eastern part of the Kemijärvi complex, northern Finland. In: Evins, P. & Laajoki, K. (eds.), Archaean and early Proterozoic (Karelian) evolution of the Kainuu-Peräpohjola area, northern Finland; a guidebook for the Nordic research field seminar organized by the universities of Oslo, Oulu and Turku, June 2-10, 1997. *Res Terrae, Ser.* A13, 66-71.

Evins, P., Laajoki, K., Mansfeld, J., et al. (2000). New geochronological constraints on sedimentation, metamorphism, and magmatism in the Suomujärvi Complex, SE Lapland, Finland. *24th Nordic Geological Winter Meeting*, Trondheim, 6-9 January, 200, Abstracts, p. 65.

Eyles, N. & Eyles, C.H. (1992). Glacial depositional systems. In: Walker, R.G. & James, N.P. (eds.), Facies models: response to sea level change. St. John's, Newfoundland. *Geological Association of Canada*, 73-100.

Eyles, N. & Menzies, J. (1983). The subglacial landsystem. In: Eyles, N. (ed.), *Glacial Geology*. Pergamon press, Oxford, pp. 19-67.

Gleeson, C.F., Rampton, V.N., Thomas, R.D., et al. (1989). Effective mineral exploration for gold using geology, Quaternary geology and exploration geochemistry in areas of shallow till. *Geol. Surv. Canada*, Paper 89-20, 71-96.

Hanski, E. (1997). The Nuttio serpentinite belt, central Lapland; an example of Paleoproterozoic ophiolitic mantle rocks in Finland. In: Messiga, B. & Tribuzio, R. (eds.), *IOS International Ophiolite Symposium; from rifting to drifting in present-day and fossil ocean basins*, Pavia (Italy), September 12-23, 1995. Ofioliti 22:1, 35-46.

Hanski, E., Huhma, H. & Vaasjoki, M. (2001). Geochronology of northern Finland; a summary and discussion. In: Vaasjoki, M. (ed.), Radiometric age determinations from Finnish Lapland and their bearing on the timing of Precambrian volcano-sedimentary sequences. *Geological Survey of Finland*, Special paper 33, 255-279.

Harris, J.R., Sanborn-Barrie, M., Panagapko, D.A., et al. (2006). Gold prospectivity maps of the Red Lake greenstone belt: application of GIS technology. *Canadian Journal of Earth Sciences* 43, 865-893.

Hartikainen, A. & Nurmi, P. (1993). Till geochemistry in gold exploration in the late Archean Hattu schist belt, Ilomantsi, eastern Finland. In: Nurmi, P. A. & Sorjonen-Ward, P. (Eds.), Geological development, gold mineralization and exploration methods in the late Archean Hattu schist belt, Ilomantsi, eastern Finland. *Geological Survey of Finland*, Special Paper 17, 273-289.

Hirvas, H. (1991). Pleistocene stratigraphy of Finnish Lapland. *Bulletin of Geological Survey of Finland* 354.

Hirvas, H., Alftan, A., Pulkkinen, E., et al. (1977). Raportti malminetsintää palvelevasta maaperätutkimuksesta Pohjois-Suomessa vuosina 1972-1976. Summary: A report on glacial drift investigations for ore prospecting purposes in northern Finland 1972-1976. Geological Survey of Finland, *Report of Investigations* 19, 54 p.

Hirvas, H. & Nenonen, K. (1990). Field methods for glacial indicator tracing. In: Kujansuu, R. & Saarnisto, M. (Eds.), *Glacial Indicator Tracing*. A.A. Balkema, Rotterdam, pp. 217–247.

Hirvas, H. & Tynni, R. (1976). Tertiääristä savea Savukoskella sekä havaintoja tertiäärisistä mikrofossiileista. Summary: Tertiary clay deposit at Savukoski, Finnish Lapland, and observations of Tertiary microfossils, preliminary report. Geologi 28, 33-40.

Huhma (1986). Reworking versus juvenile additions to the continent during the Early Proterozoic Svecokarelian Orogeny in Finalnd. *Terra cognita* 6:2, p. 236.

Huhma, H., Mutanen, T. & Whitehouse, M. (2004). Oldest rocks of the Fennoscandian Shield: the 3.5 Ga Siurua trondhjemite gneiss in the Archaean Pudasjärvi Granulite Belt, Finalnd. *Geochimica et Cosmochimica Acta* 68, A754.

Hättestrand, C. (1997). Ribbed moraines in Sweden – distribution pattern and paleoglaciological implications. *Sedimentary Geology* 111, 41–56.

Kauranne, K. (1958). On prospecting for molybdenum in the basis of its dispersion in glacial till. Bulletin de la Commission Géologique de Finlande 180, 31-43.

Kesola (1991). Taka-Lapin metavulkaniitit ja niiden geologinen ympäristö; Lapin vulkaniittiprojektin raportti. Summary: Metavolcanic and associated rocks in the northernmost Lapland area, Finland; A report of the Lapland Volcanite Project. Geological Survey of Finland, *Report of Investigation* 107, 62 p.

Kesola (1995). Näätämön kartta-alueen kallioperä. Summary: Pre-Quaternary rocks of the Näätämö map-sheet area. Suomen geologinen kartta 1:100 000; kallioperäkarttojen selitykset. *Geological Survey of Finland*, 88 p.

Klassen, R.A. (2001). A Quaternary geological perspective on geochemical exploration in glaciated terrain. In: McGlenaghan, M.B., Bobrowsky, P.T., Hall, G.E.M. & Cook, S.J. (eds.), *Drift exploration in glaciated terrain. Geological Society Special Publication* 185, 1-17. The Geological Society Publishing House, Bath.

Kleman, J. (1994). Preservation of landforms under ice sheets and ice caps. *Geomorphology* 9, 19-32.

Kleman, J. & Borgström, I. & Hättestrand, C. (1994). Evidence for a relict glacial landscape in Quebec-Labrador. *Palaeogeography, Palaeoclimatology, Palaeoecology* 111, 217-228.

Kleman, J. & Borgström, I. (1996). Reconstruction on palaeo-ice sheets: The use of geomorphological data. *Earth Surface Processes and Landforms* 21, 893-909.

Kleman, J. & Hättestrand, C. (1999). Frozen-bed Fennoscandian and Laurentide ice sheets during the Last Glacial Maximum, *Nature* 402, 63-66.

Kleman, J., Hättestrand, C., Borgström, I., et al. (1997). Fennoscandian palaeoglaciology reconstructed using a glacial geological inversion model. *Journal of Glaciology* 43, 283-289.

Koivisto, T. (1992). Discovery of a Cu-Au mineralization by step-by-step proceeding till geochemical study in humid hilly peneplain. *In:* Kauranne, K., Salminen, R., Eriksson, K. (Eds.), *Handbook of exploration geochemistry, vol. 5; Regolith exploration geochemistry in arctic and temperate terrains*. Elsevier, Netherlands, 302-310.

Koljonen, T. (1992). Results of the mapping. In: T. Koljonen (Ed.), The geochemical atlas of Finland, Part 2: Till. *Geological Survey of Finland*, pp. 106-125.

Korja, T., Tuisku, P., Pernu, T., et al. (1996). Field, petrophysiacl and carbon istobe studies on the Lapland Granulite Belt; implications for deep continental crust. *Terra Nova* 8:1, 48-58.

Korsman, K., Koistinen, T., Kohonen, J., et al. (1997). Bedrock map of Finland 1:1 000 000. *Geological Survey of Finland*.

Kujansuu, R. (1976). Glaciogeological surveys for ore prospecting purposes in Northern Finland. In: Legget, R.F. (ed.), Glacial Till. *The Royal Society of Canada, Special publication* 12, 225-239.

Larson, P.C. & Mooers, H.D. (2005). Generation of a heavy-mineral glacial indicator dispersal train from a diabase sill, Nipigon region, northwest Ontario. *Canadian Journal of Earth Science* 42, 1601-1613.

Lehtonen, M.L., Marmo, J.S., Nissinen, A.J., et al. (2005). Glacial dispersal studies using indicator minerals and till geochemistry around two eastern Finland kimberlites. *Journal of Geochemical Exploration* 87, 19-43.

Lehtonen, M., Airo, M.L., Eilu, P., et al. (1998). Kittilän vihreäkivialueen geologia; Lapin vulkaniittiprojektin raportti. Summary: The stratigraphy, petrology and geochemistry of the Kittilä greenstone area, northern Finland; a report of the Lapland Volcanite Project. Geological Survey of Finland, *Report of Investigation* 140, 144 p.

Marker, M. (1988). Early Proterozoic thrusting of the Lapland Granulite Belt and its geotectonic evolution, northern Baltic Shield. *Geologiska Föreningens i Stockholm Förhandlinga*r 110:4, 405-410.

Menzies, J. & Shilts, W.W. (1996). Subglacial environments. In: J. Menzies (Ed.), Past *Glacial Environments; Sediments, forms and techniques* (Glacial Environments Series; Vol. 2). Butterworth-Heinemann, pp. 15-136.

McGlenaghan, M. (2005). Indicator mineral methods in mineral exploration. Geochemistry: Exploration, *Environment, Analysis* 5, 233-245.

Mutanen, T. & Huhma, H. (2003). The 3.5 Ga Siurua trondhjeimite gneiss in the Archaean Pudasjärvi Granulite Belt, northern Finland. *Bulletin of the Geological Society of Finland* 75, 51-68.

Mäkinen, K. (2005). Dating the Weichselian deposits of southwestern Finnish Lapland. In: Ojala, A. E. K. (ed.), Quaternary studies in the northern and Arctic regions of Finland, Proceedings of the workshop organized within the Finnish National Committee for Quaternary Research (INQUA), Kilpisjärvi Biological Station, Finland, January 13-14th 2005. *Geological Survey of Finland*. Special Paper 40, 67–78.

Nykänen, V. (2006). Spatial analysis as prospectivity mapping tool. Geological Survey of Finland 1/12/2006. *http://en.gtk.fi/Research/CR/Spatial_analysis.html*.

Nykänen, V.M. & Salmirinne, H. (2007). Prospectivity analysis of gold using regional geophysical and geochemical data from the Central Lapland Greenstone Belt, Finland In: Ojala V.J. (ed.), Gold in the Central Lapland Greenstone Belt. *Geological Survey of Finland*, Special Paper 44, 235-253.

Nykänen, V.M, Ojala, V.J., Sarapää, O., et al. (2007). Spatial modelling techniques and data integration using GIS for target scale gold exploration in Finland. In: Milkereit, B (ed.), Exploration in the new millenium; *Proceedings of the Fifth Decennial International Conference on Mineral Exploration*. Toronto, 911-917.

Perttunen, V. & Vaasjoki, M. (2001). U-Pb geochronology of the Peräpohja Schist Belt, northwestern Finland. In: Vaasjoki, M. (ed.), Radiometric age determinations from Finnish Lapland and their bearing on the timing of Precambrian volcano-sedimentary sequences. *Geological Survey of Finland*, Special Paper 33, 45-84.

Peuraniemi, V. (1982). Geochemistry of till and mode of occurrence of metals in some moraine types in Finland. *Geological Survey of Finland*, Bulletin 322, 75 p.

Peuraniemi, V. (1990a). The weathering crust in Finnish Lapland and its influence on the composition of glacial deposits. In: Aario, R. (ed.), Glacial heritage of Northern Finland; an excursion guide. Nordia tiedonantoja, *Sarja A*, 1990, 1, 7-11.

Peuraniemi, V. (1990b). Heavy minerals in glacial material. in: R. Kujansuu and M. Saarnisto (eds.), *Glacial indicator tracing*. A. A. Balkema, Rotterdam, 165-185.

Piotrowski, J. (ed.) (1997). Subglacial environments. *Sedimentary Geology* 111:1-4. 330 p.

Pulkkinen, E., Keinänen, V., Sarala, P., et al. (2007). Heavy minerals in gold exploration at Petäjälehto, northern Finland. In: *23rd International Applied Geochemistry Symposium (IAGS), exploring our environment*, Oviedo, Spain, 14-19 June 2007. Extended abstracts. Oviedo, University of Oviedo, 1-8. Optical disc (CD-ROM).

Rastas, P., Huhma, H., Hanski, E., et al. (2001). U-Pb isotopic studies on the Kittilä greenstone area, central Lapland, Finland. In: Vaasjoki, M. (ed.), Radiometric age determinations from Finnish Lapland and their bearing on the timing of Precambrian volcano-sedimentary sequences. *Geological Survey of Finland*, Special Paper 33, 95-141.

Räsänen, J. & Huhma, H. (2001). U-Pb datings in the Sodankylä schist area, central Finnish Lapland. In: Vaasjoki, M. (ed.), Radiometric age determinations from Finnish Lapland and their bearing on the timing of Precambrian volcano-sedimentary sequences. *Geological Survey of Finland*, Special Paper 33, 153-188.

Räsänen, J. & Vaasjoki, M. (2001). The U-Pb age of felsic gneiss in the Kuusamo schist area: reappraisal of local lithostratigraphy and possible regional correlations. In: Vaasjoki, M. (ed.), Radiometric age determinations from Finnish Lapland and their bearing on the timing of Precambrian volcano-sedimentary sequences. *Geological Survey of Finland*, Special Paper 33, 143-152.

Saarnisto, M. (2005). Rannansiirtyminen ja maankohoaminen; Itämeren vaiheet ja jokien kehitys. In: Johansson, P. & Kujansuu, R. (eds.), Pohjois-Suomen maaperä – Maaperäkarttojen 1:400 000 selitys. Summary: Quaternary deposits of Northern Finland – Explanation to the maps of Quaternary deposits 1:400 000. *Geological Survey of Finland*, Espoo, 164-170. (in Finnish with English summary)

Saarnisto, M. & Lunkka, J.P., (2004). Climate variability during the last interglacial-glacial cycle in NW Eurasia. In: Battarbee, R. W., Gasse, F. & Stickley, C. E. (eds.) Past climate variability through Europe and Africa. *Developments in Paleoenvironmental Research* 6, 443–464.

Saarnisto, M. & Tamminen, E. (1987). Placer gold in Finnish Lapland. In: Kujansuu, R. & Saarnisto, M. (eds.), INQUA Till Symposium, Finland 1985. *Geol. Surv. Finland*, Special Paper 3, 181-194.

Salminen, R. (Ed.) (1995). Regional geochemical mapping in Finland in 1982-1994. Geological *Survey of Finland, Report of Investigation* 130.

Salminen, R. & Tarvainen, T. (1995). Geochemical mapping and databases in Finland. *Journal of Geochemical Exploration* 55, 321-327.

Salonen, V-P. (1988). Application of glacial dynamics, genetic differentiation of glacigenic deposits and their landforms to indicator tracing in the search for ore deposits. In: Goldhtwait, R.P. & Matsch, C. (eds.), Genetics *Classification of Glacigenic Deposits*. Balkema, Rotterdam, pp. 183-190.

Salonen, V.P., Kaakinen, A., Kultti, S., et al. (in press). *Middle Weichselian glacial event in the central of the Scandinavian Ice Sheet recorded in the Hitura pit*, Ostrobothnia, Finland. Boreas.

Sarala, P. (2003). Ribbed-moreenit – jäätikön liikesuunnan poikittaiset indikaattorit. Summary: Ribbed moraines – transversal indicators of the ice flow direction. *Geologi* 55:9-10, 250-253.

Sarala, P. (2005a). Glacial morphology and dynamics with till geochemical exploration in the ribbed moraine area of Peräpohjola, Finnish Lapland. PhD thesis. *Geological Survey of Finland*, Espoo. 17 p.

Sarala, P. (2005b). Till geochemistry in the ribbed moraine area of Peräpohjola, Finland. *Applied Geochemistry* 20, 1714-1736.

Sarala, P. & Nenonen, J. (2005). Ore prospecting in the ribbed moraine area of Misi, northern Finland. In: Autio, S. (ed.), Geological Survey of Finland, Current research 2003-2004. *Geological Survey of Finland*, Special Paper 38, 25-29.

Sarala, P., Nykänen, V. Sarapää, O., et al. (2007b). Quaternary geological and till geochemical studies in verifying GIS-based prospectivity mapping in the Central Lapland Greenstone Belt, northern Finland. In: Pérez, J.L. (ed.), IAGS 2007 – *Exploring our environment, Program & Abstracts*. 23 rd IAGS, 14.-19.6.2007, Oviedo. University of Oviedo, Spain, 161-162.

Sarala, P. & Peuraniemi V. (2007). Exploration using till geochemistry and heavy minerals in the ribbed moraine area of southern Finnish Lapland. *Geochemistry: Exploration, Environment, Analysis* 7:3, 195-205.

Sarala, P., Peuraniemi, V. & Aario, R., (1998). Glacial geology and till geochemistry in ore exploration in the Tervola area, southern Finnish Lapland. *Bulletin of the Geological Society of Finland* 70, 19–41.

Sarala, P. & Rossi, S. (2000). The application of till geochemistry in exploration in the Rogen moraine area at Petäjävaara, northern Finland. *Journal of Geochemical Exploration* 68, 87-104.

Sarala, P. & Rossi, S. (2006). Rovaniemen - Tervolan alueen glasiaalimorfologiset ja - stratigrafiset tutkimukset ja niiden soveltaminen geokemialliseen malminetsintään. Summary: Glacial geological and stratigraphical studies with applied geochemical exploration in the area of Rovaniemi and Tervola, southern Finnish Lapland. *Geol. Surv. Finland, Rep. Invest.* 161.

Sarala, P., Rossi S., Peuraniemi V., et al. (2007a). Distinguishing glaciogenic deposits in southern Finnish Lapland: implications for exploration. *Applied Earth Science* (Trans. Inst. Min. Metall. B) 116:1, 22-36.

Sarapää, O. (1996). Genesis and age of the Virtasalmi kaolin deposits, southeastern Finland. In: Sarapää,l O. (ed.), Proterozoic primary kaolin deposits at Virtasalmi, southeastern Finland; 1996, *Geol. Surv. Finland*, Espoo, pp. 87-152.

Sauramo, M. (1924). Tracing of glacial boulders and its application in prospecting. Bulletin de la Commission *Géologique de Finlande* 67, 37 p.

Shilts, W.W. (1972). Drift prospecting; geochemistry of eskers and till in permanently frozen terrain, District of Keewatin, N.W.T. *Geological Survey of Canada*, Paper 72-45, 34 p.

Shilts, W.W. (1975). Principles of geochemical exploration for sulphide deposits using shallow samples of glacial drift. *The Canadian Mining and Metallurgical Bulletin* 68:757, 73-80.

Silvennoinen, A. (1991). Kuusamon ja Rukatunturin kartta-alueiden kallioperä. Summary: Pre-Quaternary rocks of the Kuusamo and Rukatunturi map-sheet areas. Suomen geologinen kartta 1:100 000, kallioperäkarttojen selitykset lehdet 4524+4542, 4613. *Geological Survey of Finland*, 62 p.

Sturt, B.A., Melezhik, V.A. & Ramsay, D.M. (1994). Early Proterozoic regolith at Pasvik, NE Norway: palaeoenvironmental implications for the Baltic Shield. *Terra Nova* 6, 618-633.

Sutinen, R. (1992). Glacial deposits, their electrical properties and surveying by image interpretation and ground penetrating radar. *Bulletin of the Geological Survey of Finland*, 359 p.

Svendsen, J., Alexanderson, H., Astakhov, V., et al. (2004). Late Quaternary ice sheet history of northern Eurasia. *Quaternary Science Reviews* 23, 1229–1271.

Söderman, G. (1985). Planation and weathering in eastern Fennoscandia. *Fennia* 163, 347-352.

Tuisku, P., Mikkola, P. & Huhma, H. (2006). Evolution of migmatic granulite complexes: implications from Lapland Granulite Belt, Part 1: Metamorphic geology. *Bulletin of the Geological Society of Finland* 78:1, 75-105.

Vuollo, J. & Huhma, H. (2005). Paleoproterozoic mafic dikes in NE Finland. In: Lehtinen, M., Nurmi, P. & Rämö, O.T. (eds.), *Precambrian geology of Finland: key to the evolution of the Fennoscandian Shield. Developments in Precambrian geology* 14, Amsterdam, Elsevier, 195-236.

Väänänen, J. (2004). Sieppijärven ja Pasmajärven kartta-alueiden kallioperä. Summary: Pre-Quaternary rocks of the Sieppijärvi and Pasmajärvi map-sheet areas. Suomen geologinen kartta 1:100 000, kallioperäkarttojen selitykset lehdet 2624 ja 2642. *Geological Survey of Finland*, 55 p.

Wennervirta, H. (1968). Application of geochemical methods to regional prospecting in Finland. *Bulletin de la Commission Géologique de Finlande* 234, 91 p.

Chapter 2

GEOCHEMICAL MODELING OF CONCENTRATED MINE WATERS: A COMPARISON OF THE PITZER ION-INTERACTION THEORY WITH THE ION-ASSOCIATION MODEL FOR THE STUDY OF MELANTERITE SOLUBILITY IN SAN TELMO MINE (HUELVA, SPAIN)

Javier Sánchez España[1]* and Marta Diez Ercilla[1]
[1]Instituto Geológico y Minero de España, 28003, Madrid, Spain

ABSTRACT

This work describes recent hydrogeochemical research carried out in an extremely acidic (pH 0.61-0.82) hypersaline brine (TDS=225 g/L, including 74 g/L (1.33 M) Fe, 134 g/L (1.39 M) SO_4^{2-}, 7.5 g/L (0.28 M) Al, 3 g/L (0.12 M) Mg, 2 g/L (0.03 M) Cu, and 1 g/L (0.015 M) Zn) which seeps from a pyrite pile and eventually forms evaporative pools of green, mostly ferrous, ultra-concentrated water in the massive sulphide mine of San Telmo (Huelva, Iberian Pyrite Belt, SW Spain). The physico-chemical conditions of this water occasionally allow the formation of attractive crystals of Zn-rich melanterite ($Fe^{II}SO_4·7H_2O$), whose solubility is directly related with the Fe(II) and SO_4^{2-} aqueous concentration, as well with pH and T. Geochemical modeling carried out with the PHREEQCI computer code and using the ion-association model (with the Davis equation) and the Pitzer specific-ion-interaction approach (with MacInnes convention) indicates that, although both models have limitations, the Pitzer approach describes better the water/melanterite equilibrium problem. The Davis approach can not explain the formation of this mineral at the conditions found in San Telmo, whereas the Pitzer model suggests that, at the moment of sampling, the acidic brine was very slightly subsaturated with respect melanterite (SI_{Mel}=0.0 to -0.40). The Pitzer approach, however, predicts melanterite precipitation at cool temperatures (T<2-3 °C) and/or Fe and SO_4^{2-} concentrations slightly higher than the ones measured in the water, which may perfectly account for the presence of this mineral in the studied system. Geochemical research using

---

*Corresponding author: j.sanchez@igme.es

advanced modeling tools may help to solve water/mineral equilibrium problems in solutions of extreme ionic strength such as ultra-concentrated mine waters.

Keywords: AMD, ionic strength, ion activity, Pitzer method, melanterite, pyrite oxidation, IPB.

1. INTRODUCTION AND AIMS OF THE STUDY

The geochemical study of hypersaline, concentrated brines is often complex due to the extremely high density of ions in solution. At ionic strengths above 2-3.5 mol/kg, *binary* interactions between species of like charge and *ternary* (or simultaneous) interactions between three or more ions may occur, and dilute solution concepts (in which the activities of the solutes are ideal and equal molal concentrations) do not apply for these systems (Langmuir, 1997; Nordstrom, 2004). The standard Debye-Hückel and Davies equations for dilute solutions can not satisfactorily describe the behavior of ions in these brines, and the geochemical modeling of such concentrated waters (aimed at calculating ion activities, but also predicting mineral solubilities and geochemically relevant processes such as acid-base reactions) is best accomplished with the Pitzer specific-ion-interaction model, which has been shown to accurately predict the behavior of electrolyte solutions with ionic strengths up to 6 mol/kg (Pitzer, 1986).

Extremely acidic mine drainage (AMD) waters formed after the oxidative dissolution of pyrite and other sulphides exposed to atmospheric oxygen and water in underground and surface mine workings, and present in pools or as pore waters in waste piles or tailings constitute a special case of hypersaline brine which usually show extreme acidity content (both *mineral* or *total* acidity, and *free* acidity with very high concentration/activity of protons [H^+] in solution, and therefore very low pH) and abnormally high concentrations of other solutes (such as Fe(II), Fe(III), SO_4^{2-}, and also Zn and Cu), as a result of the combination of different processes such as acid generation by pyrite oxidation, concentration of H^+ and other ions by evaporation, microbial oxidation of Fe(II) to Fe(III), and acid production during evaporative salt formation (Nordstrom et al., 2000). Good examples of such acidic liquors in mining environments have been reported in mine sites such as Iron Mountain, California (Nordstrom and Alpers, 1999a,b; Nordstrom et al., 2000), Heath Steele, Canada (Blowes et al., 1990, 1991), Sherridon, Canada (Ptacek and Blowes, 2003), or Genna Luas, Sardinia, Italy (Frau, 2000). Chemical analyses of these ultra-concentrated waters usually exhibit pH values from below 0.6-1.0 to as low as -3.6 (Nordstrom et al., 2000), and yield unusally high metal contents (e.g., 48-111 g/L Fe_{total}, 0.2-4.8 g/L Cu, 6-55 g/L Zn) and sulphate (85-760 g/L SO_4^{2-}). Despite the extremely inhospitable conditions of these systems for life growth, there exist acidophilic microorganisms (including Fe(II)-oxidizing bacteria and archaea) which not only inhabit these systems, but also appear to have a key role as controls of the aqueous composition (e.g., Edwards et al., 1999, 2000; Druschel et al., 2004; González-Toril et al., 2003).

A routinary field survey carried out in 2006 in San Telmo mine (Iberian Pyrite Belt – IPB– mining district, Huelva, Spain), resulted in the sampling and analyses of a extremely low pH (0.61-0.82) acidic leachate with very high concentrations of dissolved sulphate and metals. The physico-chemical characteristics of this leachate, which emerged from a pyritic waste pile and formed a small pool a few meters downstream, far exceeded the extremest conditions found in surface AMD waters of the IPB until that moment (see compilation in Sánchez-España et al., 2005).

Preliminary attempts to geochemically model this aqueous system using conventional geochemical software and dilute solution concepts (PHREEQCI software with the Davies and extended Debye-Hückel equations, i.e., the ion-association or ion-pairing method) failed in describing the water/mineral equilibrium of this system, as it predicted undersaturation of the water with respect to melanterite, a mineral which was observed to extensively precipitate from the water. Consequently, further modeling efforts were made by using the Pitzer theory of specific-ion-interaction for high ionic strength solutions, a model which has been successfully applied for mine-drainage waters (Alpers and Nordstrom, 1991, 1999; Ptacek and Blowes, 1994, 2000, 2003; Nordstrom et al., 2000; Frau, 2000). This paper reports the results of such geochemical calculations, also describing the chemical composition of the hyperacidic-hypersaline mine water, as well as those of the melanterite crystals formed in the water. The aim of the study was two-fold: (i) firstly, to define whether or not a water/mineral equilibrium existed in the acidic brine, also inferring the control played by melanterite solubility on the aqueous composition, and (ii) secondly, to evaluate the validity and applicability of the available models (ion-association theory with extended forms of the Debye-Hückel equation, and specific-ion-interaction model using the Pitzer equations) for calculating ion-activity coefficients and saturation indices of minerals in the studied water.

2. ENVIRONMENTAL DESCRIPTION OF THE MINE SITE

San Telmo mine is located in the northwesternmost corner of the Spanish part of the Iberian Pyrite Belt (IPB) mining district, a world-famous and giant massive sulphide province which has been extensively exploited from pre-Roman times (as soon as 2000 BC), and specially during the 20th century, for the recovery of S (basically used by the sulphuric acid production industry) and base metals like Cu, Zn, Pb, Ag and Au (Pinedo Vara, 1963; Leistel et al., 1998). Like the majority of mines in the IPB, the San Telmo ore deposit was exploited by both underground and surface mining, and included a vaste open pit which was abandoned and flooded in 1989. The mine site contains large piles of pyrite-rich wastes (some of them are 20-30 m high and hundreds of meters long) from which highly acidic leachates (pH 2-3) emerge and transport large quantities of metals, sulphate and acidity from the mine site to a local stream (Sánchez-España et al., 2005).

In addition to these large piles, which are mostly formed by accumulation of volcanic and sedimentary host rocks extracted from the pit during the mine operations, there is a pile of much smaller scale (5-6 m high, 20-30 m long) which is exclusively formed by ground pyrite accumulated for transport purposes (Figure 1A). This pyrite pile is clearly differentiated from the large-scale piles in that it usually shows widespread accumulation of turquoise-blue melanterite covering the surroundings of the pile (mostly formed by fine-grained pyritic material transported by gravity flows, wind and/or runoff during rainstorm events). The dominantly pyritic composition of this pile also determines the composition of the seepage emerging from it, which exhibits much more acidic conditions (pH 0.6-0.8) compared with the not so acidic leachates observed in the rest of piles (pH 2-3).

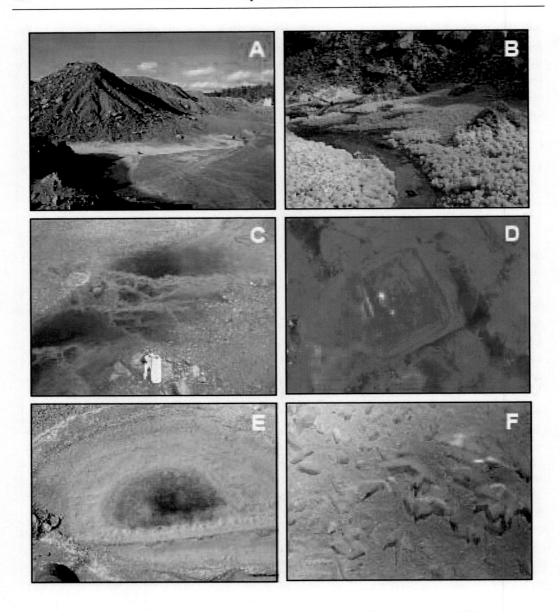

Figure 1: Melanterite occurence and field aspect of the studied site: (A) View of the sulphidic, pyrite-chalcopyrite-quartz–rich waste pile from which the acidic leachate emerges. (B) Melanterite efflorescences forming in the margins of the acidic, Fe(II)-rich leachate from capilary waters moving upwards in the pyritic sludge. (C) Micro-terracets formed by melanterite precipitation along a very slow-flow reach of the acidic leachate. (D) Melanterite crystals precipitating from a extremely acidic, Fe(II)-SO_4^{2-}-rich liquor in a melanteritic brine at the surface of the pyritic sludge. (E) View of a small acidic pool isolated in the pyritic sludge and showing abundant melanterite crystals precipitating in the water column. (F) Detail of (E). Photographs B, E and F are courtesy of Professor F. Velasco (UPV-EHU).

The acidic leachate which seeps from its base is not permanent, but a sporadic drainage which flows downslope from the pile during days or weeks after rainstorm events. This acidic leachate eventually accumulates in small depressions and forms pools of even more acidic and concentrated water in which evaporation re-concentrates all solutes and enhances the precipitation of attractive efflorescent evaporative salts (mainly melanterite ranging from turquoise-blue to emerald-green in colour) during dry episodes. The water sampled and analyzed in this study (illustrated in Figure 1) formed a small pool over the pyritic substrate after seeping from the pile, and showed a vivid emerald-green color as well as abundant crystals of melanterite growing in the aqueous medium.

3. METHODOLOGY AND ANALYTICAL TECHNIQUES

3.1. Field Measurements and Sampling

The field measurements and the sampling of water, minerals and microorganisms were carried out on March 2006, under relatively dry and cool environmental conditions (~50-70% relative air humidity and ~12-16 °C air temperature).

Field parameters (pH, Eh, temperature (T), dissolved O_2 (DO), and electric conductivity (EC)), were directly measured in the acidic waters (both in the discharge point and in the pool). These data were obtained with HANNA portable instruments (probe types HI 9025 (pH, Eh, T), HI 9145 (DO) and HI 9033 (EC)) calibrated on site against supplied calibration standards supplied by HANNA. Because the glass membrane electrode used for pH measurements could only be calibrated against 7.01 and 4.01 standard buffers (much higher than the pH of the studied solution), the error of the electrode used (which could *a priori* include acid errors and liquid-junction deviations) was evaluated against internal solutions of sulphuric acid (0.1 M H_2SO_4 and 1 M H_2SO_4 solutions). Based on the experimental findings of Nordstrom et al. (2000), these concentrated solutions should yield pH values approaching 1.0 and 0.0, respectively (so that it can be assumed ideal behaviour and m_{H+}~a_{H+}). The electrode readings were 1.03 for the 0.1 M H_2SO_4 solution and 0.20 for the 1 M H_2SO_4 solution. In accordance, an error of around ±0.1 was statistically calculated and assumed for the pH measurements.

The water samples were taken with a 60 ml-syringe, filtered on site with 0.45 μm membrane filters (Millipore), stored in acid-washed 125 ml-polyethylene bottles, and refrigerated at 4°C during transport to the laboratory.

3.2. Laboratory Analyses

Water samples were analyzed by AAS for Na, K, Mg, Ca, Fe, Cu, Mn, Zn and Al, ICP-AES for Ni, and ICP-MS for As, Cd, Co, Cr, Pb and U. ^{115}In was used as internal standard for calibration of the ICP-MS analyses. Sulphate was gravimetrically measured as $BaSO_4$. The accuracy of the analytical methods was verified against certified reference waters. The detection limits for trace elements were 10 μg/L for Zn, 2 μg/L for Ni, Co, Cr, Pb and U, 0.4 μg/L for As, Cd and Cu. The detection limit for major cations (Na, K, Ca, Mg, Mn, Fe, Al) was <1 mg/L in all cases.

Measurement of Fe(II) was performed on site by reflectometry, with a digital *Rqflex10* reflectometer (Merck) and Reflectoquant® analytical strips with 2.2'-bipiridine, after diluting the sample with acidified distilled water. The concentration of total iron was also measured by this same procedure, after reducing all dissolved ferric iron by addition of ascorbic acid ($C_6H_8O_6$) to the sample. The concentration of Fe(III) was calculated by the difference between the ferrous and total iron concentrations. When compared with a spectrometric technique, the reflectometric method is reasonably accurate (comparison of the total iron concentration measured by this method with that obtained by AAS yielded a match of around 93% between both methods), but it has the advantage of ensuring a reliable and immediate quantification of the ferrous to ferric iron ratio. Further, the theoretical Eh value calculated with PHREEQC from the Fe(II)/Fe(III) field measurements (630 mV) matches the Eh value measured in the field and corrected to the standard hydrogen electrode (626 mV).

The melanterite crystals taken from the water pool were analyzed by XRF (PHILIPS 1404) for the elements Si, Al, Fe, Ca, Mn and Mg, elemental analyzer (Eltra CS-200) for total S, ICP-MS (after digestion with HNO_3 and H_2O_2) for Cd, Co, Th, U, V and Zn, and AAS for Na, Cu, As and Pb. Certified international reference materials (BCS 175/2, BCS 378, FER-1, FER-2) were used to check the accuracy of the analytical data. The detection limits for trace elements were 10 µg/g for Ni, 2 µg/g for Co, Cu, Th, V and Zn, and 0.2 µg/g for Cd and U. The samples were mineralogically characterized by X-ray Diffraction (XRD) with a PHILIPS PW 1710 diffractometer, CuKα radiation (40 kV, 40 mA).

3.3. Geochemical Modeling

The computer codes PHRQPITZ (Plummer et al., 1988; version 1.12, released June 1994) and PHREEQCI (Parkhurst and Appelo, 1999; version 2.13.2, released February 1, 2007) were used for the geochemical simulations. The code PHRQPITZ is a FORTRAN 77 computer program that incorporates the Pitzer specific-ion-interaction theory of activity correction for calculation of activity coefficients in brines and other electrolyte solutions to high concentrations (Pitzer, 1973, 1975, 1979, 1986; Harvie and Weare, 1980; Harvie and others, 1984). This software allows to predict ion activities and mineral solubilities in highly concentrated brines with very high ionic strengths far above that of seawater (0.7 molar), giving the possibility of using either the MacInnes (1919) convention or the unscaled option for calculation of single-ion activity coefficients (Plummer et al., 1988; Nordstrom et al., 2000). The version of the PHREEQCI software used in this study incorporates several databases (MINTQ, WATEQ4F, PHREEQC, PITZER) which contain thermodynamic data for a large number of dissolved species, and allows to calculate the distribution of aqueous ionic species and their corresponding activities by either the Davies equation or the extended Debye-Hückel equation, but it also allows to calculate ion-activity coefficients and mineral saturation indices using the Pitzer equations by just selecting the PITZER thermodynamic database. For the evaluation of the melanterite solubility over a given pH and T range, the default database provided with PHRQPITZ was enlarged with thermodynamic data of this mineral given in the WATEQ4F database (log K=-2.21, $\Delta H_r°$=+4.91 kcal/mol; Ball and Nordstrom, 1991).

4. RESULTS AND DISCUSSION

4.1. Water Chemistry of the Hyperacidic Mine Water

The aqueous composition of the acidic leachate found in San Telmo, along with the composition of some other acidic mine drainage waters from the IPB, and some other concentrated acidic waters reported in the literature for mine sites in several parts of the world, is provided in Table 1.

The pH measurement of the water yielded values of 0.61±0.10 (discharge point) and 0.82±0.10 (pool), which are among the lowest pH values ever reported for a free-flowing, surface AMD water of the IPB, being also in the same order than other ultra-concentrated waters (mostly pore-waters) studied in other mine sites of the world (Table 1). At the moment of sampling (around midday, at ambient temperature of 16°C), the water presented a temperature of 20°C (although it must have been much cooler during the early morning, when the ambient temperature was as low as 2-3°C; *Red de Estaciones Agroclimáticas, Junta de Andalucía*), an electric conductivity of 125 mS/cm, an Eh value (oxidation-reduction potential referenced to the standard hydrogen electrode) of 582 mV (discharge point) and 625 mV (pool), and a virtually total anoxia with less than 0.14 mg/L of dissolved O_2 (eq. to 3% sat.).

The concentration of sulphate (134,200 mg/L SO_4^{2-}) and metals (e.g., 74,215 mg/L Fe_t, 7,556 mg/L Al, 1,945 mg/L Cu, 1,096 mg/L Zn, 3,036 mg/L Mg, 13,759 µg/L Cd) measured in the pool water are among the highest ever measured in a mine drainage water of the IPB, and comparable to values obtained in pore-waters of tailings in other mine districts (Table 1). But this abnormally metal-enriched water composition is specially obvious as regards to the trace metal content, which includes concentrations in the order of thousands of µg/L for elements which are usually very close to or below the detection limits in more standard (pH 2-3) acidic mine waters of the IPB (e.g., 9,247 µg/L Bi, 6,600 µg/L Li, 10,431 µg/L Mo, 2,683 µg/L Tl, 6,756 µg/L V, 3,268 µg/L W, 1,322 µg/L Y; Table I). Some exceptions to this pattern of extreme element concentration are the cases of Na (37 mg/L), Ca (137 mg/L), Mn (38 mg/L), As (303 µg/L), Co (2,447 µg/L), Cr (52 µg/L) or Pb (108 µg/L), which are not especially enriched with respect to more standard acidic drainages of the IPB (Table 1).

Special attention has to be paid on the iron speciation in the pool, which contains around a 59% of Fe(II) (40,600 mg/L) and a corresponding 41% of Fe(III) (28,630 mg/L). Together with the strong depletion in dissolved oxygen recognized in the water, the observed redox state of the Fe(II)/Fe(III) pair indicates that, despite the low pH of the water, oxidation of Fe(II) does occur in this water. These data of Fe(II)/Fe(III) concentrations measured in the brine pool on March 2006 compare well with determinations carried out on the same leachate but in the discharge point at the base of the pyrite pile (Figure 1B) in moments where no pool existed in the area. These determinations of the iron speciation (e.g., 48,900 mg/L Fe(II) (81%) *vs.* 11,700 mg/L Fe(III) (19%) on Febrary 2007) suggest that, with respect to the free-moving water segment of the acidic leachate, oxidation proceeds for a longer time in the pools where the water accumulates. Further, the isolation of small water volumes in the pools enhances evaporation which in turn leads to evapoconcentration of solutes, including sulphate and iron. This dual mechanism of oxidation plus evaporation would thus account for the apparent difference between the iron concentration and speciation detected in the pool with respect to that measured in the discharge point of the acidic leachate.

Table 1: Water composition of the hyperacidic brine found in San Telmo mine (March 2006, in bold) in the hydrochemical context of the IPB acidic mine drainage (AMD) systems ([a], average chemical composition of 62 AMD waters from 25 mines of the Iberian Pyrite Belt, taken from Sánchez-España et al., 2005; [b], average of 22 acidic mine pit lakes (AML) from the IPB, taken from Sánchez-España et al., *in press*), and other ultra-concentrated AMD waters reported in the literature (Corta Atalaya acidic pit lake, IPB, taken from Sánchez-España et al., *in press* [b]; Iron Mountain, taken from Alpers et al., 2003 [c]; Heath Steele, Eastern Canada, taken from Blowes et al., 1991 [d], Sherridon, Northern Canada, taken from Ptacek and Blowes, 2003 [e], and Genna Luas, Sardinia, Italy, taken from Frau, 2000 [f]). Abbreviations: S.U., Standard Units; u.b.d.l., usually below detection limit; (1) Measured on site by colorimetric reflectance photometry; (2) Total iron measured by AAS.

Parameter	Units	San Telmo Brine		AMD IPB[a]	AML IPB[b]	Other ultra-concentrated AMD waters				
		Discharge Feb 2007	Pool March 2006			Corta Atalaya[b]	Iron Mountain[c]	Heath Steele[d]	Sherridon[e]	Genna Luas[f]
pH	S.U.	0.61	0.82	2.7	2.9	1.2	-2.5	1	0.62	0.67
EC	mS/cm	125	-	7.9	8	55.6	-	-	-	-
Eh	mV	582	625	627	754	584	-	-	-	620
DO	mg/L	0.14	-	3.8	6.9	0	-	-	-	-
DO	% sat.	3	-	42	84	0	-	-	-	-
SO$_4$	mg/L	-	134200	7460	7900	41900	760000	85000	280000	203000
Na	mg/L	-	37	32	37	78	416	-	-	20
K	mg/L	-	130	3	3	12	194	-	-	87
Mg	mg/L	-	3036	414	433	1957	437	-	-	3615
Ca	mg/L	-	137	162	209	286	279	-	-	438
Fe$_t$	mg/L	60600(1)	74215(2)	1494	2899	36675	111000	48000	129000	77250
Fe(II)	mg/L	48900(1)	40600(1)	801	2189	32500	34500	-	-	72615
Fe(III)	mg/L	11700(1)	28630(1)	476	998	3950	76500	-	-	4635
Al	mg/L	-	7556	386	304	1919	1420	-	-	2550
Mn	mg/L	-	38	37	61	128	23	-	-	263
Cu	mg/L	-	1945	64	114	1350	4760	600	1600	220
Zn	mg/L	-	1096	169	455	6670	23500	6000	55000	10800
As	µg/L	-	303	2123	9386	158730	340000	-	50000	70000
Ba	µg/L	-	1547	7	14	-	200	-	-	-
Be	µg/L	-	190	u.b.d.l.	10	-	200	-	-	-
Bi	µg/L	-	9247	u.b.d.l.	55	-	-	-	-	-
Cd	µg/L	-	13759	490	1137	18020	211000	-	100000	60000
Ce	µg/L	-	467	u.b.d.l.	267	-	-	-	-	-
Co	µg/L	-	2447	3413	3208	18689	5300	-	-	15000
Cr	µg/L	-	52	118	137	1295	600	-	-	2721
La	µg/L	-	279	u.b.d.l.	109	-	-	-	-	-
Li	µg/L	-	6600	u.b.d.l.	139	-	-	-	-	-
Mo	µg/L	-	10431	u.b.d.l.	10	-	4200	-	-	494
Ni	µg/L	-	3220	1063	1495	5214	3700	-	-	27000
Pb	µg/L	-	108	61	494	5402	11900	-	-	7907
Rb	µg/L	-	767	u.b.d.l.	-	-	-	-	-	-
Sb	µg/L	-	599	u.b.d.l.	3	-	-	-	-	1729
Sc	µg/L	-	320	u.b.d.l.	18	-	-	-	-	-
Sr	µg/L	-	398	u.b.d.l.	248	-	-	-	-	-
Th	µg/L	-	376	23	6	-	-	-	-	-
Tl	µg/L	-	2683	27	4	-	390	-	-	1967
U	µg/L	-	981	64	7	-	-	-	-	859
V	µg/L	-	6756	65	27	-	15000	-	-	4776
W	µg/L	-	3268	u.b.d.l.	35	-	-	-	-	-
Y	µg/L	-	1322	u.b.d.l.	206	-	-	-	-	-

4.2. Chemical Composition of Melanterite in the Acidic Pool

Melanterite is present as various forms in the studied mine site, including: (i) micro-crystals formed onto the surfaces of the pyrite grains within the pile, (ii) efflorescences formed in the margins of the acidic leachate in the base of the pile (Figure 1A-B), (iii) small crystal aggregates precipitating on the surface of the pyritic sludge, (iv) small terracets formed by micro-crystals growing transversely to the flow direction in the slow flowing acidic leachate (Figure 1C), and (v) large crystals directly precipitated from the water in small pools after accumulation and evaporation of the acidic water (Figure 1D-F). The mineralogical identification and chemical composition given below refers to this last type of melanterite.

Table 2: Chemical analyses of the melanterite crystals precipitating from the acidic mine water found in San Telmo

Element	Concentration	Units
SiO_2	0,22	%
Al_2O_3	0,26	%
FeO	31,56	%
CaO	0,2	%
MnO	0,05	%
MgO	0,85	%
LOI	68,8	%
SO_3	37,25	%
Ba	12	ppm
Cd	7	ppm
Co	186	ppm
Cu	7427	ppm
Pb	25	ppm
Sb	2	ppm
W	10	ppm
Zn	12557	ppm

Data from Febrary 2007.

The XRD analyses confirmed the essentially melanteritic nature of the crystals sampled from the hyperacidic pool water. However, melanterite is very unstable under anhydrous conditions and was fastly dehydrated and transformed into rozenite (thus lossing three water molecules from its stoichiometric formula) during sample preparation for determination of its chemical composition by means of XRF, ICP-MS and AAS, so that the chemical analyses provided in Table 2I indicates essentially the composition of a rozenite (based on the iron content -31.56% FeO- and the LOI -68.8%, including 37.3% SO_3 and 31.5% H_2O-). Some impurities of SiO_2, MgO, MnO, CaO and Al_2O_3 (in amounts ranging from 0.05 % wt. to 0.85 % wt.) are also observed, probably reflecting some partial cationic substitution of Fe^{2+} by other divalent metal cations like Mg^{2+} in the melanterite crystal lattice. However, the most streaking feature of the Fe(II)-salt composition is the high content of Zn (12,557 ppm) and, in a lesser extent, Cu (7,427 ppm), which collectively account for a nearly 2% wt. of the chemical formula (Table II). Considering that only Zn^{2+}, Cu^{2+} and Mg^{2+} substitute part of the Fe^{2+} in the melanterite composition, the chemical analyses gives a stoichiometric formula of $(Fe_{0.915}Zn_{0.042}Cu_{0.025}Mg_{0.017})SO_4 \cdot 7H_2O$ which is practically identical to compositions previously reported for melanterites forming in other mine drainage settings like Iron Mountain, California (Alpers et al., 1994) or Genna Luas, Sardinia (Frau, 2000).

4.3. Geochemical Modeling

4.3.1. Aqueous Chemistry and Geochemical Modeling of Extremely Low Ph Mine Drainage Environments: Previous Works in Similar Systems

Previous studies on the geochemistry and mineralogy of similarly acidic and metal-laden mine waters include the pioneering works of D.K. Nordstrom and co-workers in the Richmond mine of Iron Mountain, California (e.g., Alpers et al., 1994, 2003; Alpers and Nordstrom, 1999; Nordstrom and Alpers, 1999a,b; Nordstrom et al., 2000), but also studies carried out in the Canadian deposits of Heath Steele, New Brunswick (Blowes et al., 1991) and Sherridon mine, Manitoba (Ptacek and Blowes, 2003), and in Genna Luas, Sardinia, Italy (Frau, 2000). A compilation of geochemical and mineralogical data from these studies can be found in Ptacek and Blowes (2003), and the composition of some of these extreme mine drainage waters is given for comparison in Table I along with the San Telmo mine solution.

Among these very low pH, ultra-concentrated mine drainage solutions, only the water studied by Frau in Genna Luas, Sardinia, is comparable to the water studied in this work in terms of environmental setting (surface water at ambient conditions and seeping from a pyrite pile at an old abandoned mine), and melanterite occurrence, as the rest of ultra-concentrated waters were basically porewaters taken from tailings (e.g., Heath Steele and Sherridon) or waters coming from drippings from Zn-Cu melanterite stalactites in open raises or stopes, and from pools collecting these drips (e.g., Iron Mountain).

As regards to the geochemical modeling of these waters, Alpers et al. (1994) and Nordstrom et al. (2000) successfully used a revised version of the PHRQPITZ code (with added temperature corrections for the $Fe(II)-SO_4-HSO_4$ system over a temperature range of 25-47°C) for calculation of the H^+-ion activities and mineral equilibria in twelve waters from the Richmond mine with pH mostly in the range of 0.52 to -3.6 (the most acidic waters known to date). These authors prepared standards of sulphuric acid (H_2SO_4) with different molality (from around 0.1 to around 10) and defined pH for calibration of glass membrane electrodes using the Pitzer model with MacInnes scaling, and also made laboratory experiments to assess the effect of $FeSO_4$ addition on the response of the pH electrode to sulphuric acid solutions. Nordstrom and colleagues concluded that the addition of Fe(II) to the system had a minimal effect on the calculated hydrogen-ion activities. Also, these researchers found numerous efflorescent salts in these extreme waters (mainly melanterite, römerite and rhomboclase), and stated that such extreme waters were mainly the result of pyrite oxidation and concentration by evaporation, with minor effects from Fe(II) oxidation and efflorescent salt formation.

At Sherridon mine, Manitoba, mineral saturation indices were calculated for tailings porewater using the PHRQPITZ computer code (Plummer et al. 1988) with additional ion-interaction coefficients added by Ptacek and Blowes (1994). The model calculations indicated that the porewater was at saturation with respect to melanterite and gypsum at depths where these minerals were observed (Ptacek and Blowes, 2003). These authors noted that the ion-interaction approach seemed to provide more reliable geochemical results than the ones obtained with the ion-association model.

Blowes et al. (1991) applied a modified version of PHRQPITZ (Plummer et al. 1988) to calculate mineral saturation indices for a range of minerals (mainly gypsum and melanterite) for the Heath Steele tailings. These calculations were in agreement with mineralogical studies, but additional calculations performed with other geochemical codes like MINTEQA2 (Allison et al., 1991) and the WATE4F database (Ball and Nordstrom, 1991) gave inconsistent results as regards to the gypsum precipitation rate.

Frau (2000) also applied the original version of PHRQPITZ (Plummer et al., 1988) to calculate the saturation index for melanterite and gypsum in the highly concentrated mine drainage waters at the Genna Luas minesite, which showed a chemical composition very similar to the one found in San Telmo. The calculations indicated slightly saturated conditions with respect to melanterite and gypsum. The fact that this mineral was effectively precipitating in the site lead Frau to conclude that the slight saturation was probably associated with kinetic limitations on the precipitation rate, the presence of impurities in the mineral, or a limitation of the original PHRQPITZ database. Additional calculation carried out using PHREEQE (Parkhurst et al., 1980) and SOLMINEQQ.88 with the Pitzer option (Kharaka et al., 1988) indicated always undersaturated conditions which were inconsistent with the field observations.

In the Iberian Pyrite Belt mining district, many evaporative sulphate salts have been reported from mine sites and from the margins of rivers severely affected by AMD pollution such as the Tinto and Odiel rivers (e.g., García García, 1996; Hudson-Edwards et al., 1999; Buckby et al., 2003; Velasco et al., 2005). However, to date no attempt had been made to geochemically study and model the chemical composition of the waters from which these sulphates were precipitated. Buckby et al. (2003) reported the presence of melanterite in the banks of the Tinto river which had apparently precipitated from a water draining a pyritic waste pile. These authors reported a pH value of 0.37 for this water, although unfortunately, they did not provide any information about the procedures and methodology used for the measurement of this pH value, nor about the pH definition and calibration used for the glass-membrane electrode. Velasco et al. (2005) also reported melanterite and rozenite forming directly from pyrite oxidation in a dried pool adjacent to a waste pile in the San Miguel mine, although these authors did not include water analyses in their study.

4.3.2. The Pitzer Ion-Interaction Approach vs. the Ion-Association Model

The aim of the geochemical simulations with the PHRQPITZ and PHREEQC codes was to calculate the saturation index of the acidic water with respect to melanterite over the range of T, pH, and Fe^{2+}-SO_4^{2-} concentration considered as representative of the conditions found in the San Telmo mine site. Because the saturation index (SI) is defined as SI=log [IAP/K_{sp}], where IAP is the ion activity product, and K_{sp} is the solubility product constant, and as far as ion activities largely depend on the mathematical approach used to calculate activity coefficients of ions in solution (e.g., the extended Debye-Hückel equation, the Davis equation, the Pitzer equations, etc.), the SI value of a given aqueous solution with respect to a given mineral phase will be also dependent on the selected model, as well as on the thermodynamic database used for the calculations. Therefore, it is extremely important to ensure that the selected method for calculating ion activities is the most appropriate and theoretically reasonable option for the studied system.

For calculation of activity coefficients (γ) of dissolved solutes, both the ion-association model (with extended forms of the Debye-Hückel equation or the Davis equation) and the specific-ion-interaction model (i.e., the Pitzer approach) use the concept of ionic strength ($I=1/2\sum m_i z_i^2$, where m is molal concentration and z is the ionic charge of the ion i; Lewis and Randall, 1921). A significant difference between both models, however, is that the Debye-Hückel equation considers long-range electrostatic effects in dilute solutions (Debye and Hückel, 1923), whereas the Pitzer model also takes into account short range interactions in concentrated brines (electrolyte solutions), for which it uses virial coefficient formulation and requires interaction parameters involving the aqueous species of interest in the water (Pitzer, 1973). The mathematical method is notably different from one model to the other, with the Debye-Hückel and Davis equations being relatively simpler than the complex formulation used in the Pitzer method (Harvie et al., 1984; Plummer et al., 1988; Langmuir, 1997; Nordstrom, 2004).

Another important difference between both models is that the Pitzer approach uses the concept of *total* or *stoichiometric* strength (I_s), considering all elements as free ions in solution (e.g., Fe^{2+}, Mg^{2+}, Ca^{2+}, SO_4^{2-}) and ignoring ion pairing, whereas the ion-association (or ion-pairing) model takes into account the formation of ion pairs and complexes in solution (e.g., $FeSO_4^+$, $Al(SO_4)_2^-$, $ZnSO_4°$, $CuSO_4°$, $MgSO_4°$), and thus calculates and uses the *effective* ionic strength (I_e). Further, in comparison with the limitation of the Pitzer database (which only includes thermodynamic data and virial coefficients for a few species like Fe^{2+}, Ca^{2+}, Mg^{2+}, Mn^{2+}, Ba^{2+}, Sr^{2+}, Na^+, K^+, Li^+, B^{3+}, Cl^-, SO_4^{2-} and CO_3^{2-}), the thermodynamic databases aviable for the Debye-Hückel and Davis equations (e.g., MINTQ, WATEQ4F, PHREEQC) are much larger and contain thermodynamic data for a higher amount of chemical species.

The formation of ion pairs and complexes removes ionic charge from the solutions, so that I_e will be always lower than I_s (Langmuir, 1997). As an example, the ionic strength calculated with the Pitzer approach for the San Telmo hyperacidic water is 3.95 mol/kg. This value is clearly higher than the one calculated using the ion-association model (2.95 mol/kg). However, the value of 3.95 molar is still lower (less than a half) than the manually calculated ionic strength of 8.17 mol/kg (considering all elements present as free ions in solution), which is explained by the fact that the Pitzer database is limited to the Na-K-Mg-Ca-Fe(II)-Mn-Sr-Ba-Li-Br-H-Cl-SO_4-OH-HCO_3-CO_3-CO_2-H_2O system, and thus ignores important elements in the studied AMD system such as Fe^{3+}, Al^{3+}, Cu^{2+} and Zn^{2+}, which are present in high molar concentrations (512 mM, 280 mM, 31 mM and 17 mM, respectively; Table 1).

However, because by definition $a_i=\gamma_i m_i$ (where a is the ion activity, m is molar concentration and γ is the activity coefficient), the smaller ion activity coefficients computed from higher ionic strengths in the stoichiometric model, tend to be compensated for by higher ion molalities in that approach (Langmuir, 1997). This fact is well illustrated in Figure 2 for the case of SO_4^{2-}. The activity coefficient calculated by the Pitzer approach is significantly lower than the one calculated with the Davis equation (Figure 2A), but because in the Davis model the total sulphate content is strongly speciated (forming ion pairs and complexes with metal cations like Fe^{2+}, Fe^{3+}, Al^{3+}, Cu^{2+} or Zn^{2+}), the molality of the free sulphate (SO_4^{2-}) content is notably lower than in the Pitzer model (which does not consider metal-sulphate speciation; Figure 2B), so that the overall result is that the SO_4^{2-} activity calculated in the Pitzer model is higher than in the Davis model (Figure 2C). Consequently, the resultant ionic activity product will be always higher in the Pitzer approach than in the Davis model, and the corresponding Pitzer's saturation index for melanterite will be also higher than Davis's at any given condition of pH and T, as shown in Section 4.3.3.

Because the ionic strength calculated for the water with the extended Debye-Hückel and the Davis equations far exceeded the value of 0.7 molar (and therefore was too high to justify use of the extended Debye-Hückel and Davis equations; Langmuir, 1997; Nordstrom, 1999, 2004), the Pitzer equations were the preferred option in this study, using both the PHRQPITZ and the PHREEQCI computer codes. However, with the purpose of illustrating how far the ion-association model deviates from the specific ion-interaction theory in the studied aqueous system, the calculations performed with PHREEQCI v.13 using the Davis equation and the PHREEQC database are also shown in the figures provided in Section 4.3.3.

The Pitzer approach is the most accepted model for high ionic strength solutions, and has been shown to accurately model the behavior of electrolyte solutions up to 6 mol/kg (Pitzer, 1973, 1987). There exist two different methods for defining ion-activity coefficients with the Pitzer model (the "unscaled" approach, *a priori* more appropriate for very high concentrations, and the "MacInnes scaled" option, which is simpler and more flexible for a wide compositional range; Alpers et al., 2003). In the PHRQPITZ code, it is possible to choose between the MacInnes and the unscaled options, whereas the PHREEQCI code uses by default the MacInnes convention. Because some slight differences (in the second or third decimal) were detected between the activity coefficients calculated by PHRQPITZ (working in MS DOS) and those of PHREEQCI (working in MS Windows) depending on the processor used, and as far as no solid criterion was found by which to select one model/code or the other, the Pitzer calculations were initially performed by the three different methods (PHRQPITZ with the MacInnes convention, PHRQPITZ with the unscaled option, and PHREEQCI v.13 with the MacInnes option and the Pitzer database). A revision of the complex formulation and equations used for calculation of activity coefficients used in the Pitzer approach can be found in Pitzer's original papers (Pitzer, 1973, 1975, 1979, 1986), in the works of Harvie and Weare (1980) and Harvie et al. (1984), and in the user's guide of the PHRQPITZ computer code (Plummer et al., 1988), and extensive discussions about the application of the Pitzer theory for geochemical calculations in brines are available in textbooks such as Langmuir (1997) or Millero (2001), or in review papers such as those of Nordstrom (1999, 2004) or Ptacek and Blowes (2003).

4.3.3. Effects of Ph, T, [Fe^{2+}] and [SO$_4^{2-}$] on the Melanterite Solubility

The results of the geochemical simulations are illustrated in Figure 3 (temperature-dependence of melanterite solubility), Figure 4 (pH-dependence of melanterite solubility), and Figure 5 (dependence of melanterite solubility on the Fe(II) and SO_4^{2-} concentration).

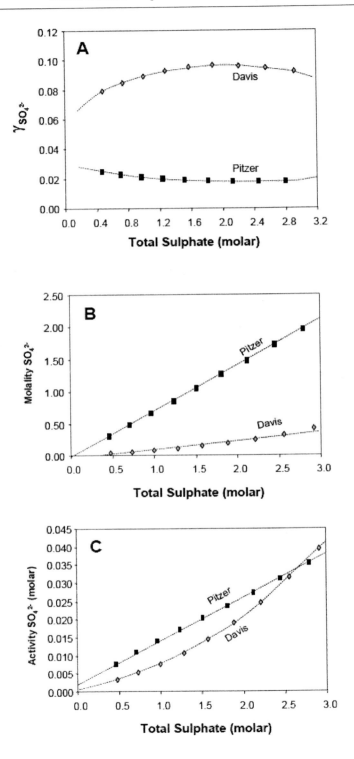

Figure 2: Variation of the SO_4^{2-} activity coefficient ($\gamma_{SO_4^{2-}}$, A), SO_4^{2-} molality (B), and SO_4^{2-} activity (C) with respect to the total sulphate concentration, as computed with the Davis and the Pitzer approaches for conditions of pH 0.6, T=20°C and [Fe(II)]=0.727 M.

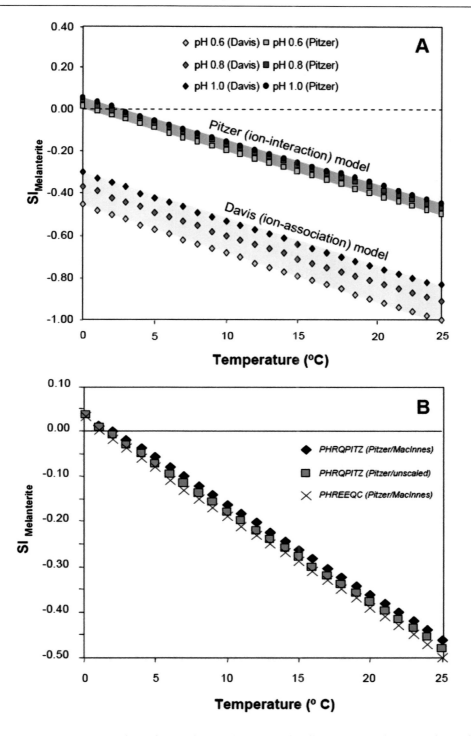

Figure 3: (A) Temperature-dependence of the melanterite solubility (given by the Saturation Index, SI_{Mel}) for different pH values (pH 0.6, pH 0.8, and pH 1.0). (B) Comparison of the calculations made with PHREEQC v.13 (taken from A) with those carried out with PHRQPITZ using both the unscaled and the MacInnes options. Calculations performed with PHREEQC v.13 (Davis and Pitzer/MacInnes options) for concentrations of 0.727 molar (40.6 g/L) Fe(II) and 1.39 molar (134 g/L) SO_4^{2-}.

In Figure 3, the effect of water temperature on the solubility of melanterite is given for several pH values (pH=0.6, value measured at the discharge point; pH=0.8, value measured in the acidic pool; pH=1.0, value measured in waters adjacent to the pool). The solubility of melanterite is here illustrated by its saturation index (SI_{Mel}) as calculated with the Pitzer and Davis equations (Fig. 3A). A positive saturation index ($SI_{Mel}>0$) denotes that the water is saturated with respect to melanterite, and therefore that there exists thermodynamic tendency for precipitation, whereas a negative saturation index indicates undersaturation and tendency for dissolution. When $SI_{Mel}=0$ then acidic solution and melanterite are said to be under quemical equilibrium (IAP=K_{sp}). The calculations were carried out for 0.73 molar Fe(II) (40.6 g/L) and 1.4 molar SO_4^{2-} (134 g/L), which are the conditions found in the acidic pool at the moment of sampling (Table 1).

The most evident conclusion which stands out from Fig. 3A is that, regardless of the pH value selected, the formation of melanterite in the acidic water is always favoured at lower ambient temperatures, while higher temperatures tend to enhance its dissolution. These results are in good agreement with the field observations (melanterite crystals are mainly observed to form and persist in the pools of the mine site during the winter, when ambient temperatures are cooler), and reflect the fact that most minerals (and melanterite in particular) dissolve endothermically ($\Delta H_r^\circ>0$; for melanterite $\Delta H_r^\circ=+4.91$ kcal/mol) and thus increase in solubility with increased temperature (Langmuir, 1997). Although it is also common to observe melanterite efflorescences around pyrite piles during the summer at higher ambient temperatures, these usually form by evaporation processes from pore and interstitial waters which are probably more concentrated than the acidic water analyzed in San Telmo.

A second conclusion emerging from Figure 3A is that, as discussed earlier, the Davis equation seems to be of less applicability for concentrated brines such as the one studied in San Telmo. The Davis model always predicts negative SI_{Mel} values at any given temperature in the range 0-25°C, which obviously desagrees with the mineralogical findings. On the other hand, the Pitzer model allows the precipitation of melanterite for T<2-3°C. This value, however, still is notably lower than the T measured in the water at the moment of sampling (20°C), which may suggest that melanterite could have precipitated during the night or the early morning, when ambient temperatures approached 0°C (*Red de Estaciones Agroclimáticas, Junta de Andalucía*). Alternatively, it could also indicate that melanterite could have precipitated at a Fe(II) and/or SO_4^{2-} concentration slightly higher than the ones measured in the water and used in the calculations, as discussed later.

As regards to the different options used with the Pitzer method (PHRQPITZ unscaled, PHRQPITZ MacInnes scaled, and PHREEQC MacInnes scaled), Fig. 3B shows that these options yielded very similar numbers which differ very little from one to another. The results of PHRQPITZ with the MacInnes option provided always the highest SI values, although the difference with respect to the other two options decreases and tend to disappear as T decreases and approaches 0°C. Overall, the differences observed between the different Pitzer options were considered unimportant for the purpose of this study.

Figure 4A describes the pH-dependence of melanterite solubility for the pH range of 0.0-3.0 at two different temperatures (2°C and 20°C). These temperatures were selected to cover a representative spectrum for the whole year in the mine site. Both the Pitzer and the Davis models provide SI_{Mel}-pH plots that become pH-independent (asymptotic to the X axis) at around pH 1.0 (Pitzer model) or 2.0 (Davis model). This trend simply reflects the pH-dependency of the respective molalities of the Fe^{2+} and SO_4^{2-} species calculated by the two methods (Fig. 4B). However, an important difference exists between both approaches in that the Pitzer model suggests that the Fe^{2+} content is the main factor controlling the solubility of melanterite (SO_4^{2-} molality >> Fe^{2+} molality), whereas in the Davis model, the SO_4^{2-} content appears to be the most critical factor (Fe^{2+} molality >> SO_4^{2-} molality; Figure 4B).

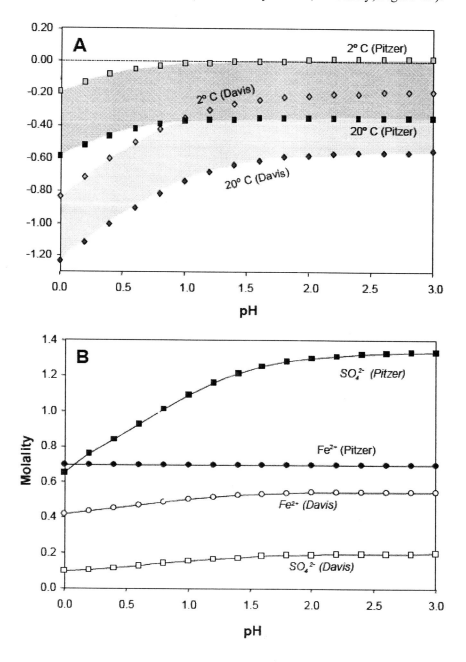

Figure 4: (A) pH-dependence of the melanterite solubility (given by the Saturation Index, SI_{Mel}) for 2 °C and 20 °C using the Davis and the Pitzer methods. (B) Variation of the Fe^{2+} and SO_4^{2-} molalities with pH. Calculations performed with PHREEQC v.13 for concentrations of 0.727 molar (40.6 g/L) Fe(II) and 1.39 molar (134 g/L) SO_4^{2-}.

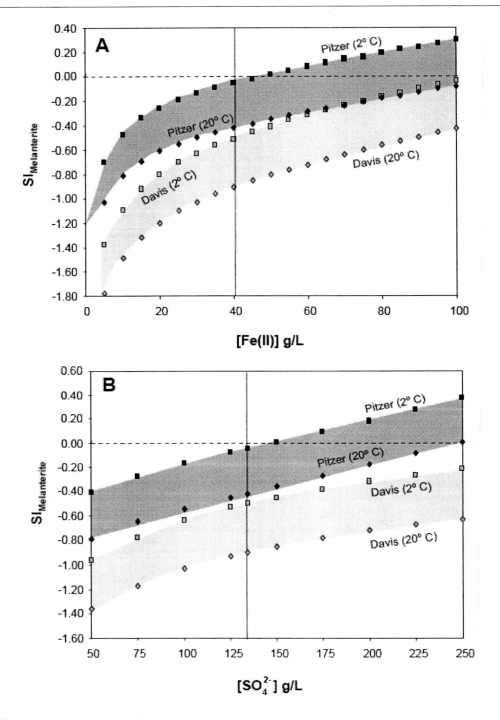

Figure 5: (A) Effect of the Fe(II) concentration on the melanterite solubility (given by the Saturation Index, SI_{Mel}) for the temperature range of 2-20 °C at pH 0.6. Calculations performed by PHREEQC v.13 with the Davis (light grey) and the Pitzer/MacInnes (dark grey) approaches, for a 1.39 M SO_4^{2-} concentration. (B) Effect of the aqueous concentration of SO_4^{2-} on the melanterite solubility in the temperature range of 2-20 °C at pH 0.6. Calculations performed by PHREEQC v.13 with the Davis (light grey) and the Pitzer/MacInnes (dark grey) approaches, for a 0.727 M Fe(II) concentration.

The Pitzer model predicts SI_{Mel} very close to 0 (and therefore indicates possible melanterite saturation/precipitation) at pH between 0.8 and 1.0 at 2°C. In accordance with Figure 3, the Pitzer model suggests that a low T (below around 2°C) is required for melanterite to precipitate (at 20°C the Pitzer's SI_{Mel} is always lower than -0.40). As in Figure 3, this fact may also suggest that the concentrations used for these calculations (0.73 molar Fe(II) and 1.4 molar SO_4^{2-}) could be slightly lower than the ones really present in the water during the precipitation process. This figure also illustrates the inadequacy of the Davis (ion-association) method in that it never allows melanterite to precipitate in the selected pH and T range.

Finally, the plots of Figure 5 illustrate the solubility control played by the Fe(II) and SO_4^{2-} concentration on the melanterite precipitation/dissolution cycle. These graphs confirm the above exposed ideas in that they: (i) indicate that the water was not at equilibrium with respect to melanterite at the moment of sampling, yielding negative SI_{Mel} values at 20°C for the concentrations of Fe(II) and SO_4^{2-} measured in the water (e.g., around 74 g/L and 134 g/L, respectively), (ii) suggest that the precipitation of melanterite was possible with the cited Fe(II) and SO_4^{2-} concentrations if T was around 2°C, and (iii) reveal the Pitzer approach as the most appropriate model for the studied water (because it allows the precipitation of melanterite at the T-pH conditions found in San Telmo). The Pitzer model predicts melanterite precipitation for Fe(II) concentrations between around 44 g/L to around 100 g/L, and for SO_4^{2-} concentrations between 140 g/L to around 250 g/L, depending on the T used in the computations. On the other hand, the Davis model does not allow melanterite to precipitate at concentrations below around 100 g/L Fe(II) and/or 325 g/L SO_4^{2-}, independently of the T used, which seem rather unrealistic for the studied water.

From the results shown in Figure 5, it could be suggested that the precipitation of melanterite took place in a moment prior to sampling when Fe(II) and SO_4^{2-} concentration were slightly higher and/or T was lower, as discussed above. In the latter case, Fe(II) would be (along with T) the most important control of the melanterite precipitation/dissolution cycle, given that it is the limiting factor of the ionic activity product (Fig. 4B). Therefore, different physico-chemical processes including oxidation, evaporation and cooling, would play an important role in the melanterite solubility control. The oxidation of Fe(II) to Fe(III) by acidophilic microorganisms tend to decrease the Fe(II) concentration and thus limit the formation of this mineral, while evaporation (which increases the Fe(II) and SO_4^{2-} concentration) and cooling, both tend to enhance melanterite precipitation.

4.3.4. Validity and Limitations of the Ion-Association and Pitzer Models

The calculations made with the Davis equation and illustrated in Figures 2 to 5, were made with the PHREEQC thermodynamic database that PHREEQCI incorporates by default, but the results can be different depending on the database used. For example, the molalities and activity coefficients calculated with the Davis equation for the Fe^{2+} and SO_4^{2-} species were distinct when using the WATEQ4F or MINTEQ databases instead of the PHREEQC database, which obviously affected to the respective activities of these ions (Table III). The ionic strength calculated for the water was also different depending on the database used. Notwithstanding, the higher Fe^{2+} activities calculated with a given database with respect to the others are compensated for by lower SO_4^{2-} activities in that database and viceversa, so that the overall result is a very similar value of the SI_{Mel} value regardless of the database used (Table 3). Moreover, the variations of the SI_{Mel} value introduced by selecting one database or the other within the ion-association model are significantly small when compared with the differences observed between the ion-association and the Pitzer models (Table 3). However, attention has to be paid in the databases used with the geochemical software packages in order to ensure internal consistency, especially when attempting to compare different sets of geochemical calculations from different sources.

Table 3: Values of ionic strength (I_s), activity of Fe^{2+} (a_{Fe2+}), activity of SO_4^{2-} (a_{SO42-}), and saturation index of melanterite (SI_{Mel}), as computed with the PHREEQC computer code and the Davis equation using different thermodynamic databases (PHREEQC, WATEQ4F, MINTEQ). The calculations obtained with the Pitzer method are also shown for comparison. Calculations in all cases were obtained for conditions of pH=0.6, T=20°C, Fe(II)=0.727 molar and SO_4^{2-}=1.39 molar.

	PHREEQC	WATEQ4F	MINTEQ	PITZER
I_s	2,95	2,06	3,32	3,95
a_{Fe2+}	0,075	0,33	0,1	0,143
a_{SO42-}	0,012	0,0036	0,0097	0,018
SI_{Mel}	-0,89	-0,76	-0,87	-0,42

The comparisons made between the results of the Pitzer approach with those obtained with the Davis equation must be taken with caution. The thermodynamic database used with the Davis equation is much larger than the limited Pitzer database (in which virial coefficients for ionic interactions have been only defined for a few species). Therefore, in the Davis approach it is possible to introduce all the chemical species analyzed in the water, whereas in the Pitzer method only a few elements could be considered (Fe(II), K, Na, Ca, Ba, Sr, Li, B, Mg, Mn, and sulphate), and some major elements like Fe(III), Al, Cu and Zn were ignored in the Pitzer computations. The inclusion of these elements would increase the total ionic strength and would also modify the activity coefficients of Fe^{2+} and SO_4^{2-}, and thus, the final result of the Fe^{2+} and SO_4^{2-} activities and the corresponding SI_{Mel} value. These variations, however, would affect in different ways to the different terms of the Pitzer model equations, which are linear algebraic functions of $\ln \gamma$ that contain binary virial coefficients describing the interaction of species of opposite sign (which are function of ionic strength) and ternary virial coefficients that account for interactions among two like-charged and a third unlike-charged species (and are independent of ionic strength; Langmuir, 1997). Therefore, it is difficult to ascertain the effect that a more complete data base with virial coefficients for Fe(III), Al, Cu, and Zn would have on the calculations of the SI_{Mel} value.

Some limitations of the Pitzer model must be also considered as regards to the thermodynamic data used for melanterite. The thermodynamic properties used in the databases are defined for pure phases, whereas the melanterite analyzed in this study contained a significant amount of impurities like Zn^{2+} or Cu^{2+} in the cristalline lattice that could introduce slight deviations from its theoretical behavior.

Additional computations using the Geochemical Workbench™ (Bethke, 1998) and the chemical data reported in Table I, have confirmed the results obtained in this study with the Pitzer method and using the PHREEQCI and PHRQPITZ computer codes. This program failed in describing the mineralogical observations with the Debye-Hückel option (regardless of the Fe(II) concentration and T used), but it satisfactorily explained the formation of melanterite at aqueous Fe(II) concentration above 66 g/L when water evaporation and pyrite oxidation were allowed to proceed simultaneously (Prof. F. Velasco, Basque Country University; *written communication*). This program, however, has probed to be more accurate with the database "*thermo.com.v8.r6*" (compiled by Wolery and Daveler, 1992) rather than with the database "*thermo.Pitzer*".

In any case, as stated by Nordstrom (1999) "...A difficulty with SI values arises when attempting to define how close to zero is close enough to say the water is in equilibrium with respect to a given mineral, and a certain margin of error must be allowed to account for errors in the water analyses and errors in the thermodynamic data...". However, the fact the water composition reflects the same molar proportion of Fe to SO_4^{2-} than melanterite (nearly 1:1, very unusual in most mine waters of the IPB with higher pH and more oxidizing conditions; Sánchez-España et al., 2005), and the observation that well developed melanterite crystals were present in the pool, support the idea that the composition of this brine is strongly influenced by the dissolution/precipitation cycle of melanterite. This mineral can thus act as a buffering system for the aqueous solution, tending to precipitate when the Fe(II) and/or SO_4^{2-} concentrations in the water exceed the equilibrium conditions (for example by evapoconcentration), and dissolving when these elements are decreased in concentration (for example by dilution during a rainstorm event).

5. CONCLUSION AND OUTLOOK

This work has provided hydrogeochemical data for extremely acidic and ultraconcentrated water seeping from a pyrite pile in San Telmo mine. The use of the Pitzer specific-ion-interaction model for calculation of ion activities indicates that this water was close to (but not at) equilibrium (it showed a slight undersaturation) with respect to melanterite at the moment of sampling. It is considered that melanterite precipitated from the acidic water at lower temperature (e.g., during the night or the early morning) and/or at slightly higher Fe(II) and SO_4^{2-} concentrations than the ones measured in the water.

Both the ion-association model (with either the Davies and/or extended Debye-Hückel equations) and the specific-ion-interaction (Pitzer) approach present limitations for calculation of activity coefficients and saturation indices of minerals in solutions with very high ionic strength (the first one does not consider short-termed ionic interactions between ions of the same sign, when such interaction do take place in highly concentrated (electrolytic) solutions, whereas the second one still has a limited thermodynamic database which does not consider the existence of different ionic pairs and complexes nor important cations like Fe(III), Al^{3+}, Zn^{2+} or Cu^{2+}). However, the Pitzer model seems to be more coherent and to better describe the mineralogical findings. The relative complexity of the mathematical expressions used by this model in comparison with the extended Debye-Hückel or the Davis equations is currently solved by the available computer codes (e.g., PHREEQCI). Future efforts in defining binary and ternary virial coefficients for more cations and incorporating them to the thermodynamic databases available for the Pitzer method will surely improve the applicability and reliability of the Pitzer method for concentrated mine waters. Further, an interesting option for geochemical modeling of concentrated brines is to use hybrid methods that take advantage of the best aspects of both approaches (for example, by determining the metal speciation with the Debye-Hückel or Davis equations, and then calculating activity coefficients and mineral solubilities with the Pitzer approach).

The use of advanced modeling tools combined with high quality hydrochemical and mineralogical data and an appropriate thermodynamic database may help to solve water/mineral interaction and equilibrium problems in solutions of high ionic strength such as mine waters or industrial wastewaters. An interesting option is to combine approach in which the speciation.

ACKNOWLEDGMENTS

Jesús Reyes and Juan Antonio Martín Rubí (IGME) are thanked for their laboratory analyses of waters and solids. This manuscript benefitted from discussions with Prof. Francisco Velasco (Basque Country University), who is also acknowledged for providing some pictures.

6. REFERENCES

Allison, J.D., Brown, D.S., & Novo-Gradac, K.J. (1991). MINTEQA2/PRODEFA2, A geochemical assessment model for environmental systems, Version 3.0 User's Manual, US Environ. Prot. Agency (EPA/600/3-91/021).

Alpers, C.N. & Nordstrom, D.K. (1991). Evolution of extremely acid mine waters at Iron Mountain, California –Are there any lower limits to pH?; in *Proceedings, 2nd International Conference on the abatement of acidic drainage, MEND (Mine Environment Neutral Drainage)*, Ottawa, Canada, v. 2, 321-342.

Alpers, C.N. & Nordstrom, D.K. (1999). Geochemical modeling of water-rock interactions in mining environments. In: Plumlee, G.S., and Logsdon, M.J. (eds.), The Environmental Geochemistry of Mineral Deposits, Part A. Processes, Techniques, and Health Issues: Society of Economic Geologists, *Rev. Econ. Geo.*, 6A, 289-323.

Alpers, C.N., Nordstrom, D.K. & Spitzley, J. (2003). Extreme acid mine drainage from a pyritic massive sulphide deposit: The Iron Mountain end-member. *In*: J.L. Jambor, D.W. Blowes, and A.I.M. Ritchie (Eds.) *Environmental Aspects of Mine wastes*, Mineralogical Association of Canada, Short Course Series Volume 31 (R. Raeside, ed.), Vancouver, British Columbia, 2003, pp. 407-430.

Alpers, C.N., Nordstrom, D.K. & Thompson, J.M. (1994). Seasonal variations of Zn/Cu ratios in acid mine water from Iron Mountain, California. In: Environmental Geochemistry of sulphide oxidation (C.N. Alpers, Blowes, D.W., eds.). *Am. Chem. Soc. Symp. Ser.* 550, 324-344.

Ball, J.W. & Nordstrom, D.K. (1991). User's manual for WATEQ4F, with revised thermodynamic data base and test cases for calculating speciation of major, trace, and redox elements in natural waters. *US Geol. Surv. Open-file report 91*-183, 189 pp.

Blowes, D.W. & Jambor, J.L. (1990). The pore-water geochemistry and the mineralogy of the vadose zone of sulphide tailings, Waite Amulet, Quebec, Canada. *Appl. Geochem.*, 5, 327-346.

Blowes, D.W., Reardon, E.J., Jambor, J.L., et al. (1991). The formation and potential importance of cemented layers in inactive sulphide mine tailings. *Geochim. Cosmochim. Acta*, 55, 965-978.

Buckby, T., Black, S., Coleman, M.L., et al. (2003). Fe-sulphate-rich evaporative mineral precipitates from the Río Tinto, southwest Spain. *Mineral. Mag.*, 67, 263-278.

Debye, P. & Hückel, E., (1923). On the theory of electrolytes. *Phys. Z.*, 24, 185-208, 305-325.

Druschel, G.K., Baker, B.J., Gihring, T.M., et al. (2004). Acid mine drainage biogeochemistry at Iron Mountain, California. *Geochem Transactions*, 5-2, 13-32.

Edwards, K.J., Gihring, T.M. & Banfield, J.F. (1999). Seasonal variations in microbial populations and environmental conditions at an extreme acid mine drainage environment. *Appl. Environ. Microbiol.* 65, 3627-3632.

Edwards, K.J., Bond, P.L., Gihring, T.M., et al. (2000). An archaeal Fe-oxidizing extreme acidophile important in acid mine drainage. *Science* 287, 1796-1799.

Frau, F. (2000). The formation-dissolutionprecipitation cycle of melanterite at the abandoned pyrite mine of Genna Luas in Sardinia, Italy: environmental implications. *Mineral. Mag.*, 64, 995-1006.

García García, G. (1996). The Rio Tinto Mines, Huelva, Spain. *The Mineralogical Record*, 27, 275–285.

González-Toril, E., Llobet-Brossa, E., Casamayor, E.O., et al. (2003). Microbial ecology of an extreme acidic environment, the Tinto River. *Appl. Environ. Microbiol.*, 6, 4853-4865.

Harvie, C.E., Moller, N. & Weare, J.H. (1984). The prediction of mineral solubilities in natural waters: The Na-K-Mg-Ca-H-Cl-SO4-OH-HCO$_3^-$-CO$_3^{2-}$-CO$_2$-H$_2$O system to high ionic strengths at 25°C. *Geochim. Cosmochim. Acta*, 48, 723-751.

Harvie, C.E. & Weare, J.H. (1980). The prediction of mineral solubilities in natural waters: The Na-K-Mg-Ca-Cl-SO$_4$-H$_2$O system from zero to high concentration at 25°C. *Geochim. Cosmochim. Acta*, 44, 981-997.

Kharaka, Y.K., Gunter, W.D., Aggarwal, P.K., et al. (1988). SOLMINEQ.88: A computer program for cgeochemical modeling of water-rock interaction. *US Geol. Surv. Water-Resour. Invest. Report 88-4227*, 420 pp.

Hudson-Edwards, K., Schell, C. & Macklin, M.G. (1999). Mineralogy and geochemistry of alluvium contaminated by metal mining in the Río Tinto area, southwest Spain. *Appl. Geochem.*, 14, 1015-1030.

Langmuir D. (1997). Aqueous environmental geochemistry. Prentice-Hall, Inc. Upper Saddel River, NJ, 602 p.

Leistel, J.M., Marcoux, E., Thiéblemont, D., et al. (1998). The volcanic-hosted massive sulphide deposits of the Iberian Pyrite Belt. Review and preface to the Thematic Issue. *Mineral. Deposita*, 33, 2-30.

Lewis, G.N. & Randall, M. (1921). The activity coefficient of strong electrolytes. *J. Am. Cem. Soc.*, 43, 1112-1153.

MacInnes, D.A. (1919). The activities of the ions of strong electrolytes. *Contrib. Research Lab. Physical Chemistry Massachusetts Institute of Technology*, 115, 1086-1092.

Millero, F.J. (2001). *The physical chemistry of natural waters*. Wiley, NY, 654 pp.

Nordstrom, D.K. (1999). Some fundamentals of aqueous geochemistry. In: Plumlee, G.S., and Logsdon, M.J. (eds.), The Environmental Geochemistry of Mineral Deposits, Part A. Processes, Techniques, and Health Issues: Society of Economic Geologists, *Rev. Econ. Geo.*, 6A, 117-123.

Nordstrom, D.K. (2003). Effects of microbiological and geochemical interactions in mine drainage. *In*: J.L. Jambor, D.W. Blowes, and A.I.M. Ritchie (Eds.) *Environmental Aspects of Mine wastes, Mineralogical Association of Canada, Short Course Series Volume 31* (R. Raeside, ed.), Vancouver, British Columbia, 2003, pp. 227-238.

Nordstrom, D.K. (2004). Modeling low-temperature geochemical processes, *In*: *Treatise on Geochemistry*, H.D. Holland and K.K. Turekian, ex. Eds.: Vol. 5, Surface and Ground Water, Weathering, and Soils, J.I. Drever, ed., Elsevier Pergamon, Amsterdam, 37-72.

Nordstrom, D.K. & Alpers, C.N. (1999a). Negative pH, efflorescent mineralogy, and consequences for environmental restoration at the Iron Mountain Superfund site, California. *Proc. Natl. Acad. Sci.*, 96, 3455-3462.

Nordstrom, D.K., Alpers, C.N. (1999b). Geochemistry of acid mine waters. In: Plumlee, G.S., and Logsdon, M.J. (eds.), The Environmental Geochemistry of Mineral Deposits, Part A. Processes, Techniques, and Health Issues: Society of Economic Geologists, *Rev. Econ. Geo.*, 6A, 133-156.

Nordstrom, D.K., Alpers, C.N., Ptacek, C.J., et al. (2000). Negative pH and extremely acidic mine waters from Iron Mountain, California. *Environ. Sci. Technol.*, 34, 254-258.

Parkhurst, D.L. & Appelo, C.A.J. (1999). User's guide to PHREEQC (Version 2) – A computer program for speciation, batch-reaction, one-dimensional transport, and inverse geochemical calculations. *US Geol. Surv. Water-Resour. Investig. Rep.* 99-4259, Denver-Colorado.

Parkhurst, D.L., Plummer, L.N. & Thorstenson, D.C. (1980). PHREEQE – A computer program for geochemical calculations. *US Geol. Surv. Water-Resour., Invest. Report* 80-96, 195 pp.

Pitzer, K.S. (1973). Thermodynamics of electrolytes, 1. Theoretical basis and general equations. *Journal of Physical Chemistry*, 77-2, 268-277.

Pitzer, K.S. (1975). Thermodynamics of electrolytes, 5. Effects of higher-order electrostatic terms. *Journal of Solution Chemistry*, 4-3, 249-265.

Pitzer, K.S. (1979). Theory: Ion interaction approach, *In* R.M. Pytkowicz, (ed.), *Activity coefficients in electrolyte solutions*, v. 1, CRC Press, Inc., Boca Raton, Florida, 157-208.

Pitzer, K.S. (1986). Theoretical considerations of solubility with emphasis on mixed aqueous electrolytes. *Pure and Applied Chemistry*, 58-12, 1599-1610.

Plummer, L.N., Parkhurst, D.L., Fleming, G.W., et al. (1988). A computer program incorporating Pitzer's equations for calculating of geochemical reactions in brines. *U.S. Geol. Survey, Water-Resources Investigations Report 88-4153*, Reston, Virginia.

Ptacek, C.J. & Blowes, D.W. (1994). Influence of siderite on the pore-water chemistry of inactive mine-tailings impoundments. In: Environmental Geochemistry of sulphide oxidation (C.N. Alpers, Blowes, D.W., eds.). *Am. Chem. Soc. Symp. Ser.* 550, 172-189.

Ptacek, C.J. & Blowes, D.W. (2000). Prediction of sulphate mineral solubility in concentrated waters. In: Sulphate minerals: Crystallography, Geochemistry, and Environmental

Significance (C.N. Alpers, J.L. Jambor., D.K., Nordstrom, eds.). *Rev. Mineral. Geochem.*, 40, 513-540.

Ptacek, C.J. & Blowes, D.W. (2003). Geochemistry of concentrated waters at mine-waste sites. *In*: J.L. Jambor, D.W. Blowes, and A.I.M. Ritchie (Eds.) *Environmental Aspects of Mine wastes, Mineralogical Association of Canada, Short Course Series Volume 31* (R. Raeside, ed.), Vancouver, British Columbia, 2003, pp. 239-252.

Sánchez-España, J., López-Pamo, E., Santofimia, E., et al. (*In press*) The acidic mine pit lakes of the Iberian Pyrite Belt: An approach to their physical limnology and hydrogeochemistry. *Applied Geochemistry*.

Sánchez-España, F.J., López Pamo, E., Santofimia, E., et al. (2005). Acid Mine Drainage in the Iberian Pyrite Belt (Odiel river watershed, Huelva, SW Spain): Geochemistry, Mineralogy and Environmental Implications. *Applied Geochemistry* 20-7, 1320-1356.

Velasco, F., Alvaro, A., Suarez, S., et al. (2005). Mapping Fe-bearing Hydrated Sulphate Minerals with Short Wave Infrared (SWIR) Spectral Analysis at San Miguel mine environment, Iberian Pyrite Belt (SW Spain). *Journal of Geochemical Exploration*, 87-2, 45-72.

In: Geochemistry Research Advances
Editor: Ólafur Stefánsson, pp. 57-92
ISBN 978-1-60456-215-6
© 2008 Nova Science Publishers, Inc

Chapter 3

GEOCHEMICAL ANOMALIES CONNECTED WITH GREAT EARTHQUAKES IN CHINA

Jianguo Du[1], Xueyun Si[2], Yuxiang Chen[1], Hong Fu[3], Chunlin Jian[4] Wensheng Guo[5]

1 Institute of Earthquake Science, China Earthquake Administration, Beijing 100036, China
2 Ningxia Bureau of Seismology, Yinchuan 750001, China
3 Yunnan Bureau of Seismology, Kunming 650041, China
4 China Earthquake Network Center, China Earthquake Administration, Beijing 100036, China
5 Inner Mongolia Bureau of Seismology, Hohhot 010050, China

ABSTRACT

The goals of this chapter are to investigate the relationship between the geochemical anomalies and great earthquakes and to distinguish seismic precursors from the abnormal phenomena in order to improve the accuracy of predicting an impending earthquake. Many devastating earthquakes occurred in China. The great earthquakes left many deaths and caused a lot of economic loss. Earthquake prediction is considered as one of the most efficient approaches to mitigate seismic hazard. Predicting earthquake mainly depends on understanding the process of earthquake generation and mechanism of seismic precursors.

The establishment and development of the seismic monitoring network in China are introduced. Plenty of novel geochemical data in China has been obtained since the late 60's of last century. The geochemical anomalies frequently occurred in the seismic zones, but rarely correlated with great earthquakes. Geochemical anomalies related to the 3 June 2007 Puer M6.4 earthquake were described. The temporal and spatial features of geochemical anomalies connected with great earthquakes varied dramatically. Combining with the geological heterogeneousness that is not well understood, it is difficult to put forward the methods to correctly identify the seismic precursors. There are a few of successful examples for predicting earthquakes based on the gaseous and hydrochemical anomalies and other precursors, but a lot of failure ones.

The geochemical anomalies related to great earthquakes can be attributed to fluid mixing and water-rock interaction. The experimental data for simulating water-rock interaction associated with formation of micro-fracture in a brittle aquifer demonstrate the two-steps model: (1) water mixed with fluids of fluid inclusions in rock resulted in soluble Cl and SO_4 approached approximately equilibrium in six hours or less; (2) dissolution predominantly controlled concentrations of other ions, resulting in the concentrations increased with increasing soaking time. A genetic model, in which the role of fluids is emphasized, is proposed based on geological and geophysical investigations and experiments of rocks at high pressure and high temperature in order to highlight the mechanism of the seismic-geochemical anomalies and process of earthquake gestation.

1. INTRODUCTION

1.1 Geochemical Singals to the Siesmic Events

Geochemical signals can often bear witness to seismic activity and other deep-earth processes [1-12]. It was reported that investigation of variation of radon concentration in association with earthquake throughout the world started in 1956 when Okabe studied the correlation between variation of radon content in atmosphere and local seismicity [13, 14]. Variations of ion concentrations in groundwater, gas compositions and isotopic ratios that were found in the epicenter areas and far from epicenter before, during and after strong earthquakes are considered to be due to the action of crustal stress related to the earthquake generation and transport of seismic wave [1-3, 15-28]. More than 50 percent of historical recorded anomalous phenomena (at least 1160 items) related to earthquakes in China are fluid precursors [29]. It was suggested that the measurement of gaseous contents were more advantageous in soil-gas than in groundwater [4, 30-33]. The global earthquake zones overlap the global geothermal zones because heat carried by the upwards migration of deep-earth fluids contributes energy for geothermal field, magma generation, volcanic eruption and earthquake generation [12]. Geochemical measurements of geothermal waters and gases appear to provide better signals for the seismic events [9, 12, 33-42]. The prior results demonstrate that hydrological and geochemical measurements on the active fault zones show better correlation with earthquakes [20, 27, 42-50]. Consequently, geochemical monitoring in seismic zone can provide better signals for earthquake activity.

1.2. Brief Review

The methodology for data analysis is important to identify anomaly and distinguish the precursory signals from various kinds of vast observed data. The observed geochemical parameters of fluids vary temporally, which is caused not only by seismic activity, but also environmental parameters, such as tide, rainfall, soil temperature, barometric pressure. It is usually difficult to distinguish the anomalies caused solely by seismic activity from those by meteorological or hydrological parameters and even factitious ones. In addition, the varying patterns differentiate from place to place, i.e. from the observatory site to site, because of the geological heterogeneousness. For example, some observed parameters decline with time in some sites, some vary periodically in other sites, and others change with spikes or oscillate frequently. Considering the temporal relationship between precursors and earthquake parameters, the geochemical precursors display the clusters in precursory time, which differs for different precursors from a few hours to a few years. A monitoring parameter may gradually exit in a given confidence interval before returning, but its departure may be rapid and irreversible. Therefore, the different statistic methods are implemented for data analysis according to the data features. Arithmetic or geometric signal definitions are commonly employed to distinguish the signals. For seismic precursory identifications, many approaches for data analysis and evaluation are employed [18, 25, 51-59].

Using the methods of the confidence intervals, time series intervention analysis, exploratory analysis and heuristics, Hartmann & Levy (2005) identified the arithmetic abnormal definitions. Deviations or discontinuities in a time series are intuitionisticly utilized for signal identification. The advanced exploratory and confirmatory statistical techniques, such as factor analysis, regression tree, can provide additional insights about the proposed anomalies related to earthquake [60]. Niazi (1985) used regression analysis to classify the earthquake precursors into 11 categories with 31 subdivisions [18]. Concentrations of gaseous components in soil often show the seasonal variations. Such periodical variation can be diminished by using regression analysis [61, 14]. The pike-like anomalies are usually recognized by using "the 2σ method", in which the anomaly thresholds are calculated with mean value and standard deviation (σ). For identifying the seismic radon precursors, Zmazek et al. (2005) used the following methods: (1) deviation of Rn concentration from the seasonal average, (2) correlation between time gradients of Rn concentration and barometric pressure, and (3) regression trees within a machine learning program. It was concluded that method (1) was much less successful in predicting anomalies caused by seismic events than methods (2) and (3). Methods (2 and 3) did not fail to observe an anomaly preceding an earthquake, but showed false seismic anomalies, the number of which is much lower with (3) than with (2) [62]. Toutain et al. (2006) proposed that a mixing function F, calculated with chemical elements, was processed with seismic energy release (Es) and effective rainfalls (R), and calculated linear impulse responses of F to Es and R [58]. The Bayes discriminant analysis was also applied to distinguish the short-term and imminent precursors for earthquake in China [63].

The relationships between fluid precursors and three parameters of earthquake (magnitude, epicenter and occurring time) are the basement of earthquake prediction. Some experiential equations were established between magnitude (M) and precursory time (T), epicentral distance (D) and precursory duration (pT) on the assumption of a homogeneous model of the earth and with a deformation (strain) of $\geq 10^{-8}$ [18, 31, 34, 41, 60, 64-66]. The results of factor analysis (with varimax rotation) for the four variables displayed that epicentral distance and magnitude are grouped on the Factor 2 whereas precursory time and precursory duration are grouped on the Factor 1. The combined correlation for geochemical precursors shows a close relationship between the precursory anomalies and the considered earthquakes (r = 0.86). The length of the precursor duration is mostly longer than the length of the precursory time, suggesting the proposed precursors are related to tectonic processes causing the referenced earthquake [60]. Moreover, Sultankhodzhayev et al. (1980) put forward the empirical formula: $\log DT = 0.63 M \pm 0.15$ [67]. Hartmann and Levy (2005) proposed a 3-D precursory boundary surface, $(T)_{max} = 6975M - 0.046D - 32.731$ [60], based on the variables magnitudes, epicentral distance and precursory time. The relationship between precursory time and magnitude of main shock is different for different groups of precursors, which can be described by a linear relation, $\log T = a + bM$, where a and b are constants. Statistical analysis indicates that relationship between M and T (in day) differs for the seismic events in the subduction zone and intraplate. For instance, experiential equation for the maximum precursory time related to intraplate earthquakes in China, is $\log_{10}^{T} = 0.38M - 0.34$, but it becomes $\log_{10}^{T} = 0.6M - 1.01$ for different discipline precursors related to others [16].

The plot of magnitude versus precursory duration scatters widely, indicating the more complex relationship. The precursory duration of radon related to great earthquakes ranges from less than one day to 500 days [60], indicating the possibility to predict exact time of an impending earthquake based upon the precursors is very small. The plots of precursory times versus epicentral distances scatter widely in a triangle region, and most of the precursory time smaller than 40 days within an epicentral distance less than 100 km, which indicates the maximum precursory times decrease with increasing epicentral distance [4, 60].

The relationship between magnitude and the epicentral distance measured in kilometers is also different from discipline to discipline. For example, an empirical formula $M = -0.87 + \log D$ was proposed by Rikitake (1987) [65], and the exponential linear equation ($D = 10^{0.43M}$) was proposed by Dobrovolsky et al. (1989) [34]. The later is frequently used to identify the possible maximum D for intraplate earthquakes. The epicetral distances of fluid geochemical precursors related to the earthquakes with $M < 8.0$ in China's mainland are usually less than 1000 km [68, 69]. Using scatter plot of the epicentral distances versus magnitudes, evaluation for the empirical equations for D and M based on the earlier reported data suggested a tendency that the larger the main shock magnitude is, the larger is the epicentral distance for which a precursor is observed [4, 65]. The number fluid anomalies related to great earthquakes in China generally decreases with increasing epicentral distance [19, 23, 68, 70]. The statistical results of the fluid anomalies related to $M \geq 7.0$ earthquakes in Xinjiang, northwestern China, and vicinity show that precursory density appear to decrease with increasing epicentral distance, i.e., the anomalies widely occurred in the region with a radius of 500 km around epicenter, but at some places in the cycle region of 500-1000 km and rarely in the region 1000 km far away from the epicenters [71]. However, the August 1999 Izmit M7.6 earthquake caused 25% coseismic increase of water flux from an artesian well in Kajaran, Armenia, at the epicenter distance of 1400 km [72]. Additionally, the 21 September 1997 Chi-Chi M7.6 earthquake associates with the anomalies of radon at the distance of about 1000 km, and those of water level with epicetral distance as far as 2060 km, which was explained by the hypotheses of crustal buckling [73]. The reason for the limitation of the empirical formula application is that tectonic and stratigraphic units are never homogeneous. For example, the geochemical anomalies related to great earthquakes in China often cluster along active faults [23, 74]. The plots of magnitudes versus precursory duration, epicentral distance display evidently regional differences [60]. Furthermore, the observation stations generally distribute heterogeneously, even in the monitoring network, which must be considered when correlations between the magnitudes, the epicentral distance, the precursory time and the precursory duration are performed. Consequently, the obtained data do not allow us to make an exact prediction for the location of an impending earthquake.

The programs of earthquake prediction started in 1960's in China, Japan, USA and former U.S.S.R. The remarkable progress has been achieved in earthquake forecasting, especially in long- and medium-term predictions, in which underground fluid monitoring plays an important role [15, 19, 68, 75- 80]. The first successful example in China is the forecasting for the 26 March 1966 Baichikou M6.2 earthquake in Ningjin country, Hebei Province according to the precursors of violent variation of water level, bubbling and muddying in well water; animal behavior anomalies; and small earthquake activity in the epicentral region. Another example is the successful prediction for the 4 February 1975 Haichen M7.3 earthquake. The main precursors were the foreshocks (more than 500 small earthquakes occurred in four days prior the main shock) and anomalies of groundwater and gasses [76]. Other successful examples of imminent forecasting in China are reported by Rikitake (1980) [16], Wang (1990) [29], and Mei et al. (1993) [77]. In addition, there are many "partially-successful" examples of earthquake predictions, i.e. one or two parameters of earthquakes were correctly predicted [81-86]. However, the precursory signals are likely to differ from earthquake to earthquake [87]. The practicality of theory predicting earthquake remains contentious [60, 88-90]. Nevertheless, this is beyond the scope of the chapter. The crucial fact is that most great earthquakes occurred in the last 30 years with no any exact prediction, indicating there is a long way for us to go to predict earthquake successfully.

The prior research results about fluid anomalies corresponding to earthquakes have been reviewed by Rikitake (1976) [15], Li et al. (1985) [17], Barsukov et al. (1985a) [1], King (1986) [30], Thomas (1988) [3], Roeloffs (1988, 1996) [91,92], Ma et al. (1982) [76], Igarashi and Wakita (1995) [93], Wyss (1997b) [94], Zhang, et al. (1988a, 1990, 1999, 2000) [81-84], Toutain & Baubron (1999) [4], Du and Kang (2000a) [22], Chen et al. (2002a, 2002b) [85,86], Hartmann and Levy (2005) [60] and King et al. (2006) [27]. The precursory monitoring network and classical fluid precursory anomalies related to great earthquakes in China and a case study of geochemical anomalies related to the 3 June 2007 Puer M6.4 earthquake are briefly described in order to investigate the relationship between the geochemical anomalies and great earthquakes and to distinguish seismic precursors from the observed abnormal variations of chemical components in underground water and gases, which favors improving earthquake prediction.

2. MONITORING NETWORK FOR EARTHQUAKE IN CHINA

2.1 Earthquake Activity

Earthquake activity and seismic disaster in China are frequent and severe. Many great earthquakes have frequently occurred in China since 1900, indicating seismic activity is severe. Earthquakes with magnitude (M) $\geq$ 6.0 occurred frequently during January 1900 to June 2007, of which 128 earthquakes are with magnitudes larger than 7.0, including nine ones with M $\geq$ 8.0 (Figure 1). The great earthquakes resulted in severe disaster and large amount of casualty. The statistics data show that 33% of great continental earthquakes in the world occurred in China, resulting in about 50% of global seismic casualty; but China just occupies about 7% of the global continent and has some 22% of the global population. Therefore, many efforts have been made in order to mitigate the seismic disaster in China. It is believed that the correct prediction for earthquake will be achieved by catching the precursory signals.

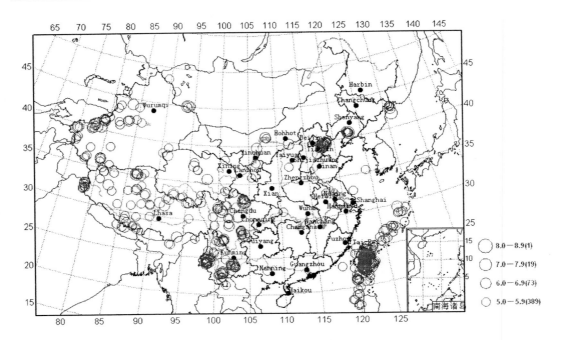

Figure 1: Epicenter distribution of earthquakes with magnitudes ≥5.0 in China from January 1900 to June 2007 (data from the Seismic Database of China Earthquake Administration).

2.2 The Monitoring Network

Underground fluid changes in many seismic regions have been observed since ancient times. It may be the earliest record in 4000 years ago that ancient Chinese wrote "Ground shocked and springs gushed when Sanmiao (a tribe) was destroyed". The famous ancient astronomer, Zhang Heng (78-139 AD), invented the first seismometer in the world and began the stage of instrumental measurement for earthquake. The emperor Kang Xi (1654-1722 AD) mentioned the active regions, sequence and generation of earthquakes in a book named as "DIZHEN" (meaning Earthquake). However, systematic monitoring such changes with the modern scientific instruments to discover precursors for earthquake did not begin until the 1960s. The construction of the monitoring network began in May 1968 in China. The monitoring network for earthquake has undergone four stages in China's mainland [22]: (1) the initial stage of the monitoring construction (1968-1975), (2) the formation stage of the network (1976-1979), (3) the regulation and reformation stage (1980-1990), and (4) the automationized stage after 1991 (Figure 2). So far, there are 493 geochemical monitoring stations for earthquake in China's mainland. Concentrations of anion and cation in waters of the wells and springs are measured at several dozens of stations, and concentrations of radon, mercury, carbon dioxide, helium etc. are monitored at others. The hydrochemical monitoring network has been well improved in the last two decades [79]. Consequently, data quality of fluid geochemistry has been improved with utilizing automatic monitoring instruments for multiple parameters.

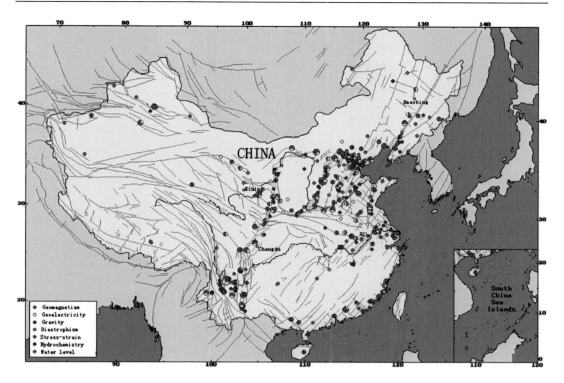

Figure 2: The monitoring network for earthquake in China's mainland. The network is consisted of 493 fluid chemistry, 504 water temperature and water level, 262 diastrophism, 55 geomagnetism, 95 geoelectricity, 14 gravity and 70 strain stations. Green lines stands for Quaternary faults.

The instrumentation for radon in the fluid geochemical monitoring network includes different types of radonscopes (e.g. FD-3017 RaA, FD-105K, SD-3A, MD-4211BA and GDK-1 Emanometers). The detection limits are less than 0.1 Bq/L, precisions are better than 10% or 15%. The different types of mercury analyzers, such as the XG-4 and RG-1 Mercury Analyzers, RG-BS Aptitude Mercury Analyzer, DFG-B Automatic Mercury Analyzer, were installed in the network with precisions of <10%. For analyzing other gaseous components, the different types of gas chromatographers (GC) were employed with precision of 5%. The hydrochemical compositions were usually analyzed by titration method and the ion chromatographers (IC) with precisions of <5%. Carbon dioxide in the fault zone was chemically determined with the glass tube in which basic chemicals were filled [95].

2.3 Some Classical Geochemical Precursors for Great Earthquakes in China

The observed geochemical data of underground fluids for 39 years show that the typical concentration anomalies of Na, Ca, Mg, Cl, F, SO_4, HCO_3 and gaseous components from the wells in China usually occurred from less than one day to several months before the earthquakes [24,25, 74, 79 81-86, 96-99]. The anomalous phenomena of underground fluids related to earthquakes can be summarized as following:

- Water level sharply increasing
- Water and/or oil pouring or erupting from the wells
- Water level violent decreasing or drying up in the wells
- Color alteration of water
- Waters becoming muddy in the well, lake and reservoir
- Bubbling or rolling in well water
- Smell alteration of water
- Springs running dry or discharge increase
- Gas eruption from the oil-well with loud sound
- Ignition above the well water and lake
- Temperature variation of well and spring waters
- Variations of radon and mercury concentrations in waters
- Variations of chemical compositions in the wells and springs

The anomalies of water level are not discussed in this chapter, but the interests are geochemical anomalies of monitored components related to great earthquakes. Some classical anomalies are shown as following.

Radon

The concentration of soil radon in a fault zone usually exhibits periodical variation, which was caused by environmental parameters. For example, radon concentrations measured in the Babaoshan fault zone in western Beijing are higher in winter season, but lower in summer [61]. Ghosh et al. (2007) reported the radon anomaly is abruptly high during rainy season that is in the month of June-July, and the radon concentration after the rainy season comes to near the same value as it was before the rainy season [14]. In this case, the seismic signals were distinguished by the regression method [61, 25, 14]. However, the concentration of soil radon in an active fault zone in western Taiwan appeared to have no periodical variation [49].

Most concentrations of radon in waters of the wells and springs (dissolved gas) varied periodically. For instance, concentrations of radon in spring water in Dingxiang county, Shanxi Province, varies periodically, which is related to flow rate on the down-flow spring (Figure 3). Others vary at irregular intervals. For example, Rn concentrations in the well water in Puer, Yunnan Province vary stochastically (Figure 4).

Radon anomalies usually occurred less six months before earthquakes [19, 85, 86, 100]. Furthermore, the long-term anomalies of radon concentrations occurred before some great earthquakes. The long-term anomalies of radon occurred 26 months before the 6 November 1988 Lanchang-Gengma M7.6 earthquake were recorded in Puer observatory at epicentral distance of 138 km, other radon anomalies at distances of 140 to 400 km were found eight months before the seismic event [25]. The increasing anomaly of Rn concentration in the well water in the Fanshan Observatory in Hebei Province, at epicentral distance of 250 km, was found before the 28 July 1976 Tangshan M7.8 earthquake (Figure 5) [101]. Radon concentrations at the Tengchong observatory, at distance of 200 km, fluctuated abnormally before the 3 June 2007 Puer M6.4 earthquake in Yunnan, resulting in both the positive and negative radon anomalies that occurred about eight months before the seismic event (Figure 6). Both positive and negative radon anomalies were found in the region of 400 km radii before the 10 January 1998 Zhangbei M6.2 earthquake [69]. Some co-seismic changes of radon concentrations in southeastern China, at epicentral distance of <550 km, related to the 21 September 1999 Chi-Chi M≥7.6 earthquake in Taiwan were reported (Figure 7) [73].

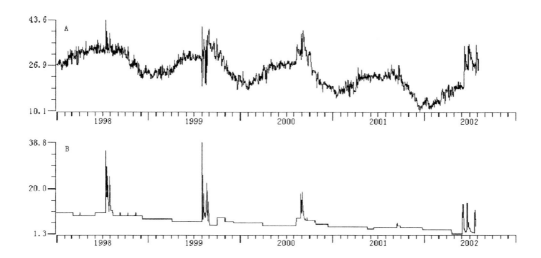

Figure 3: The time series of radon concentration (A, in Bq/L) and discharge flow rate (B, in L/min) for down-flow spring in Dingxiang County, Shanxi Province, showing annually periodical variation.

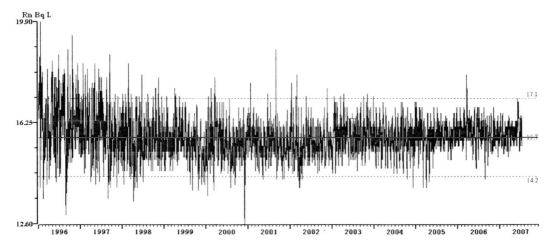

Figure 4: Time-series of daily values of radon concentrations in the well water in Puer, Yunnan Province from January 1996 to July 2007. The solid line is the mean value. The dashed lines, namely the limits of anomalies, are 2σ deviated from the mean value.

Mercury

Most concentrations of mercury in soil gas and waters seem to pulsate with time, and partially appear the periodical variations. The mercury anomalies related to great earthquakes usually occurred several months before the main shocks [25]. For instance, the mercury anomalies were found at the Midu Observatory in Yunnan Province, at <200 km epicentral distance, before three earthquakes with M>5.0 in 1995 (Figure 8). However, there is no anomaly related to the 12 July 1995 Menglian M7.3 earthquake at the observatory (D < 200 km). The systematical research on Hg anomalies and 26 earthquakes with M ≥ 5.0 in Yunnan Province and vicinity shows that the anomalies of mercury generally occurred three months before the seismic events, the precursory durations of the Hg anomalies are 20 to 90 days [102, 103]. The anomalies of mercury in the Wang-4 well in Tianjing (D = 100 km) occurred before and after the 4 July 2006 Wen'an M5.1 earthquake (Figure 9).

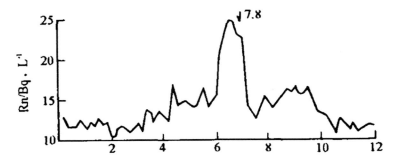

Figure 5: Variation of radon concentration from 1st January to 31 December 1976 at the Fanshan observatory (D = 250 km), showing the increasing anomaly related to the 1976 Tangshan earthquake, which occurred 43 days before the seismic event [101].

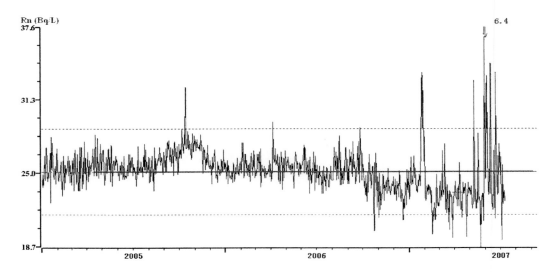

Figure 6: Diagram of Rn concentration versus time at the Tengchong Observatory (D=250 km) from January 2005 to July 2007, showing the abnormal oscillation of radon concentration, both positive and negative anomalies related to the 3 June 2007 Puer M6.4 earthquake in Yunnan Province. The solid line indicates the mean value. The dashed lines, namely the limits of anomalies, are 2σ deviated from the mean value.

Carbon Dioxide

CO_2 concentration of soil gas in the fault zones at distances less than 500 km usually increased abnormally before M>4.0 earthquakes in the capital region in China. The values of anomalies are as high as 1.3 ~ 10 times of the mean value. Precursory time is less than two months (5 ~ 47 days) (Figure 10) [104].

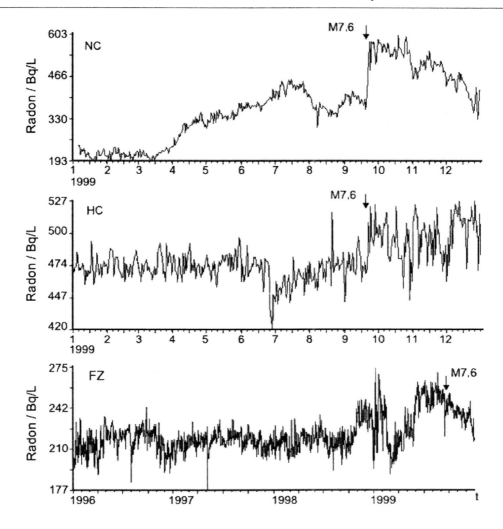

Figure 7: The temporal variations of radon at epicentral distance of <550 km, showing some typical response changes in southeastern China to the 21 September 1999 Chi-Chi M ≥ 7.6 earthquake in Taiwan, the radon precursors in the FZ well, co-seismic changes in the NC and HC wells [73].

Gases in the groundwater of five deep wells in the southern area of the Kamchatka peninsula were collected with a sampling frequency of three days for a decade. In the decade, five M>6.5 earthquakes occurred at distances less than 250 km from these wells. The hydrogeochemical data show possible precursors for the earthquakes. A total of 25 anomalies of the gas contents were identified, of which nine are successfully correlated with the earthquakes and 16 failures. The successful precursors occurred from 7 to 107 days before the earthquakes [6].

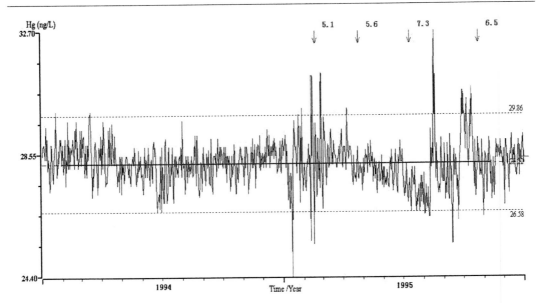

Figure 8: Time-series of mercury concentration in well water at the Midu Observatory in Yunnan Province, showing the pike-like anomalies related to the 18 February Lanchang M5.1 (D = 320 km) and 24 October 2005 Wuding M6.5 (D = 200 km) earthquakes (marked by arrows), respectively. However, there is no anomaly related to the 25 April Jingping M5.6 (D = 400 km) and 12 July Menglian M7.3 (D = 360 km) earthquakes. The dashed lines, namely the limits of anomalies, are 2σ deviated from the mean value.

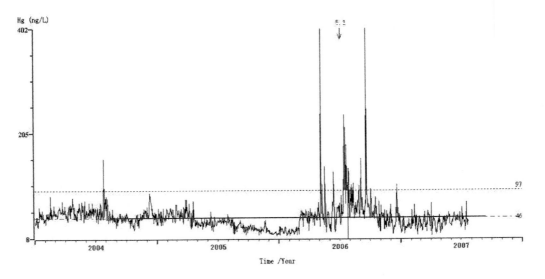

Figure 9: Time-series of Hg concentrations in water of the Wang-4 well in Tianjing, showing pike-like anomalies occurred from 4 May to 20 September and an earthquake with M5.1 occurred on 4 July 2006 (arrow).

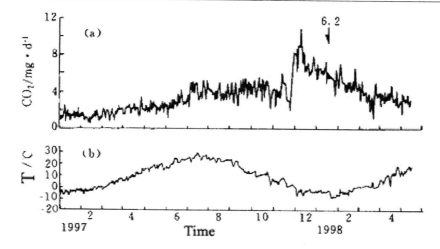

Figure 10: Concentration variation of carbon dioxide in soil gas at Houhaoyao in Hebei (D=80 km), showing precursory anomaly before the 10 January 1998 Zhangbei M6.2 earthquake [104].

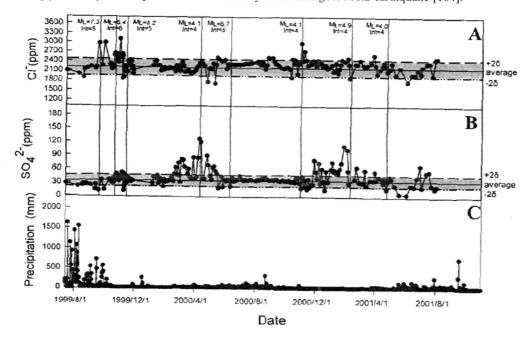

Figure 11: Temporal variations of (A) Cl^- and (B) SO_4^{2-} concentrations in the Kuantzeling hot spring in Taiwan. The vertical lines represent chemical anomalies which correlate with earthquakes that occurred in this area. (C) Daily rainfall [79].

Hydrochemical Components

The relationship between M ≥ 6.0 earthquakes (13 great earthquakes from 1988 to 2000) and hydrochemical precursory anomalies at six stations in Yunnan Province was systematically studied by Zhang and Fu (2000) [105]. The anomalies of the 19 monitored parameters such as anion, cation, pH and resistance of water, occurred often two months before the M ≥ 6.0 earthquakes. The maximum epicentral distance of the anomalies appear to be 300 km related to M ≥ 6.0 earthquakes, and 350 km related to M ≥ 7.0 earthquakes. The more than 50% anomalies of twelve variables among the 19 monitored parameters were considered to be related to the M ≥6.0 earthquakes. The resistance of water and concentrations of F^-, HCO_3^{2-}, Ca^{2+} and SO_4^{2-} seem to be more sensitive to the great earthquakes. More than 40% great earthquakes companied with the hydrocemical anomalies. It seems that there is a higher probability that another earthquake would occur if the great earthquake occurred in the precursory duration of the hydrochemistry [105]. Song et al. (2005; 2006) report that most of the hot springs and artesian springs in Taiwan clearly show chemical anomalies correlated with earthquakes. Cl^- appears to be better geochemical precursor for earthquake (Figure 11) [79, 106].

3. GEOCHEMICAL ANOMALIES RELATED TO THE 3JUNE 2007 PUER M6.4 EARTHQUAKE IN YUNNAN PROVINCE

3.1 Seimogeological Sitting

The Chuan-Dian block, in Sichuan and Yunnan Provinces, southwestern China, is tectonically located in the eastern collision zone between the India and Eurasia plates. To the east, the Chuan-Dian block is connected with the stable South China platform, and to the west, it is neighborhood with the Tibetan Plateau that is the highest mountainous region formed due to the collision between the Indian plate and the Eurasian plate. To the north, it borders with the active Kunlun and Qingling fault zones. The Chuan-Dian block is characterized by the high level of seismicity, the Quaternary volcanoes, geothermal activity and some big zones of active fault (age <100 ka), such as the Xianshuihe fault, Xiaojiang fault, and Red River fault, etc. [107-109]. Ten large earthquakes with magnitudes >7.0 occurred in this region in the last 30 years, which resulted in significant damages. The tomographic inversions revealed that P-wave velocity varies up to 7% in lithosphere in the Sichuan-Yunnan region. The velocity anomalies are attributed to enrichment of fluids, the extensional fractures of the lithosphere and/or the upward intrusion of the hot asthenospheric materials [110]. The active faults extend in three directions: N-NNW, NW and NE. The faults with NW and N-NNW trends have large lengths and are very developed (Figure 12).

3.2 Selection and Analysis of Data

Earthquake with M6.4 occurred on 3 June 2007 in Puer, Yunnan Province. There are 42 chemical stations in the monitoring network in the Chuan-Dian block, nine of which show chemical anomalies correlated to the earthquake (Figure 12). The principles for selecting and analyzing data are as the following:

1) The analyzed data were selected according to the observed parameters, data quality (reliability and continuousness) and quantity (amounts of the observed sites and observation history), the extent of understanding seismogeological background.

2) The time scale of geochemical data related to a seismic event was determined as two years before the event because most precursory time related to the great earthquakes are less than 200 days [16, 99, 105], and the maximum is 500 days [60].
3) The radius of an area, in which the observed sites for fluid geochemical parameters distribute, were selected as 500 km according to the statistical result that most epicentral distances for earthquake with M6.0 are less than 500 km and for the great earthquakes (M > 7.5) occurred in the world are usually less than 1000 km [4, 34, 60, 68, 69, 71, 99].
4) The different approaches for data analysis, such as factor analysis, "the 2σ method" and heuristics (The identification is made by breaking the normal time-series or changing sharply), were employed for abstracting the different signals, such as distinguishing anomalies, analyzing the relationship among the observed sites and parameters.

3.3 Geochemical Anomalies Related to the Main Shock

There are 42 geochemical monitoring stations of fluids in the Chuan-Dian block, and 13 geochemical anomalies are found at nine stations before and after the 3 June 2007 Puer M6.4 earthquake (Figure 12). The results of factor analysis (principle component analysis) for 32 variables show that the stations No. 22, 1, 17, 36, 19, 13, 23 and 11 contribute to the Factor 1 that reflects the constrain of active faults with NE trend and/or activity of Tengchong sub-block in the southwestern part of the Chuan-Dian block; and the stations No. 31, 34, 5 and 6 belong to the Factor 2 that reflects the affect of active faults with N-NNW trend in the eastern part of the Chuan-Dian block. The stations, at which geochemical anomalies occurred before and after the 3 June 2007 Puer earthquake, are most located in the Tengchong sub-block and related to the active faults with NE trend, indicating the earthquake and the geochemical anomalies could be caused by the same geological agent. The geochemical anomalies distribute in the region of epicentral distance from 180 to 440 km, except for one at a distance of 960 km (Table 1). However, the reason why most stations (33 stations) show no obvious anomaly in the region, especially in the epicentral area (at distance less 100 km) is not clear now. It is probably attributed to the geological complexity. Correlation between the five earthquakes with M>6.5 occurred at epicentral distances <250 km in Kamchatka from 1988 to 1997 and the hydrochemical anomalies shown that a total of 16 anomalies are with 11 successes and 5 failures. The successes appeared from 7 days to 107 days before the earthquakes. It was concluded that the relationship linking earthquakes with the hydrochemical anomalies is very complex, and no general rules for finding precursors can be assumed [5].

The features of the anomalies are shown in Table 1 and Figure 7 and 13. Most anomalies appeared one day to three months before the Puer earthquake. Rn, pH, Ca, F, Mg and HCO_3 show abnormal variations related to the Puer earthquake. The factor analysis of six variables shows that Ca, Mg, and HCO_3 belong to the Factor 1, indicating an effect of hydrochemistry. F and pH contribute to the Factor 2, indicating an effect of acidity. The hydrochemical anomalies may be mainly ascribed to contribution of fluids (gas and water) from the deep earth because F and Cl are usually derived from the deep earth in the zones of earthquake and volcano [108]. Earthquake related strain changes in the crust results into variation of pore pressure, affects fluid-rock interaction and causes the ground water to migrate in the crust, particularly along faults. Earthquake has a closely relationship with active faults. The migration of ground water and terrestrial gas emanation are mainly controlled by activities of faults and fissures in the crust [6].

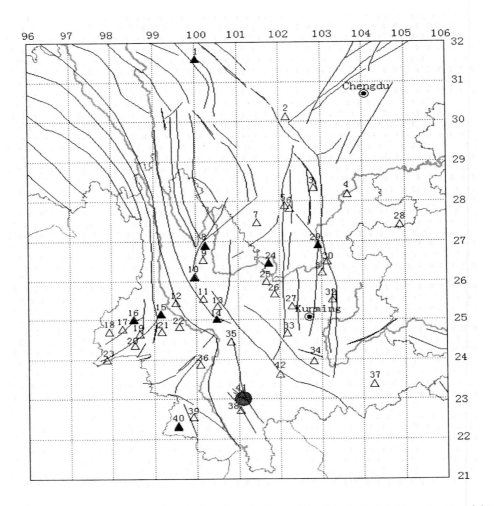

Figure 12: Distribution of the geochemical stations in Chuan-Dian block. Red dot is epicenter of the 3 June 2007 Puer M6.4 earthquake; triangles with numbers are the locations of the stations and solid triangles are the stations at which anomalies occurred 1. Ganzhi, 2. Guza, 3. Zhaojue, 4. Yongshan, 5. Taihe, 6. Xinchun, 7. Yanyuan, 8. Lijiang, 9. Heqing, 10. Eryuan, 11. Xiaguan, 12. Yongping, 13. Midu, 14. Nanjian, 15. Baoshan, 16. Tengchong, 17. Lianghe, 18. Yingjiang, 19. Longling, 20. Dehong, 21. Shidian, 22. Changhing, 23. Ruili, 24. Dukou, 25. Yongren, 26. Yuanmou, 27. Luoci, 28. Zhengxiong, 29. Qiaojia, 30. Huizhe, 31. Dongchuan, 32. Xundian, 33. Yimen, 34. Qujiang, 35. Jingdong, 36. Linchuang, 37. Wenshan, 38. Simao, 39. Lanchang, 40. Menglian, 41. Puer, and 42. Yuanjiang); dark thin line stands for fault; blue thick lines are rives.

Table 1. The features of the geochemical anomalies related to the Puer earthquake

No.	Station	Object	Epicentral distance(km)	Beginning date	Feature
1	Baoshan	Rn	310	2007.04.02	Decrease in April, dropping after the main shock
2	Baoshan	PH	310	2007.05.13	Jumping 23 days before the shock and keeping high value
3	Baoshan	Mg^{2+}	310	2006.03.12	Decreasing for 15 months and increasing 2 months before the shock,
4	Dukou	Rn	390	2007.02.06	Dropping first, then raising and reaching the highest values after the shock
5	Eryuan	F^-	360	2007.05.26	Raising one month before the shock and keeping the high values after the shock
6	Ganzhi	Rn	960	2006.12.29	Sharply dropping about 50% normal value, the shock occurred just after the anomaly ending
7	Lijiang	F^-	440	2007.06.02	Sharply raising one day before the shock and keeping high values
8	Lijiang	Ca^{2+}	440	2007.01.25	Raising for about 4 months, the shock occurred just after the anomaly ending
9	Menglian	Rn	180	2007.05.30	Sharply dropping 4 days before the shock
10	Nanjian	Rn	230	2007.06.05	The shock occurred in the day when Rn concentration started sharply dropping
11	Qiaojia	F^-	470	2007.06.24	Sharply raising after the shock
12	Tengchong	Rn	350	2007.05.09	Variation in high frequency and larger magnitude
13	Tengchong	HCO_3^{2+}	350	2007.05.30	Sharply dropping 3 days before the shock

4. MECHANISM OF GEOCHEMICAL SEIMIC PRECURSORS

Most hydrochemical variations are explained by the mixture of waters from different aquifers caused by breaking aquifuges and migration of deep fluid along active fault, and meteoric water addition to an aquifer [7, 26, 111]. For example, Koretsky (2000) summarized possible explanations for the existence of seismic precursors and their presence or absence [112]. In hypothesizing on the existence of a stress/strain transmission process, modifications of the hydrogeochemicals of water in a well may be produced by: (1) the flow into the well of waters with different chemistries, perhaps as a consequence of pumping or mixing of the output from different water-bearing strata, created by the induced stress/strain in the well zone, and (2) the circulation of the groundwater into the new zones belonging to the water-bearing stratum connected with the well, as a consequence of the intensification of the micro-fracturing processes and/or changes in existing cracks produced by the induced stress/strain. The experimental data for chemical anomalies by mixing NaCl solution to water in a hydrodynamic trench indicated that the variation amplitudes of ion concentrations attenuated obviously; and the forms varied with increasing migration distance of the solution, which was also related with the stress state [113]. However, some hydrochemical variations occurred before earthquakes can not be explained by the mixing model [114]. Before the occurrence of an earthquake, when tectonic stress increases, formation of micro-fractures in rocks could cause an increase in the surface area of rocks. As a result, radon concentration rises in the groundwater [35, 75].

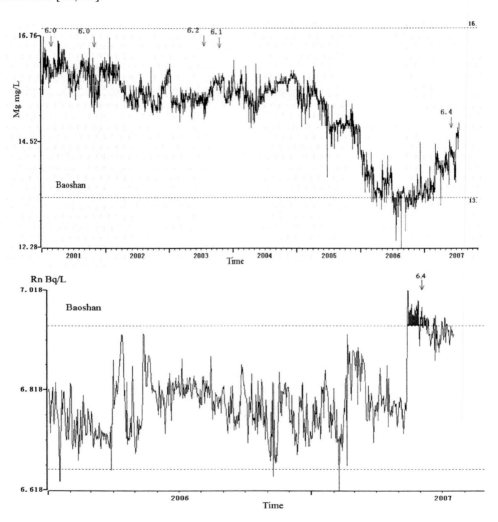

Figure 13 Continued on next page.

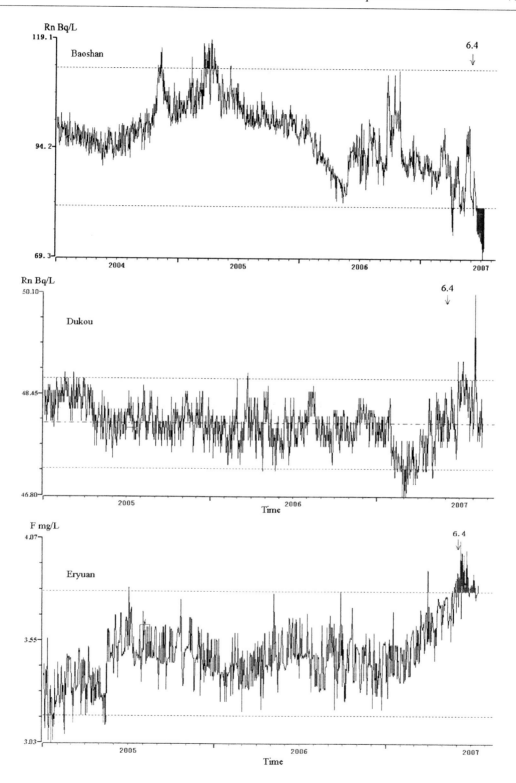

Figure 13 Continued on next page.

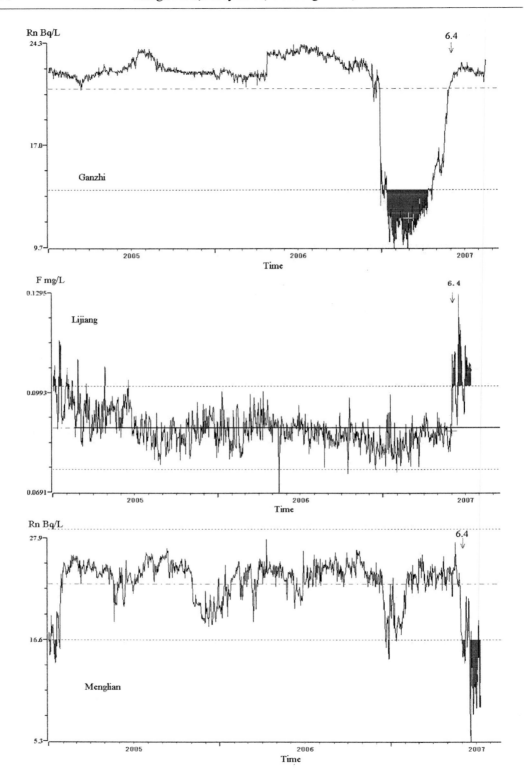

Figure 13 Continued on next page.

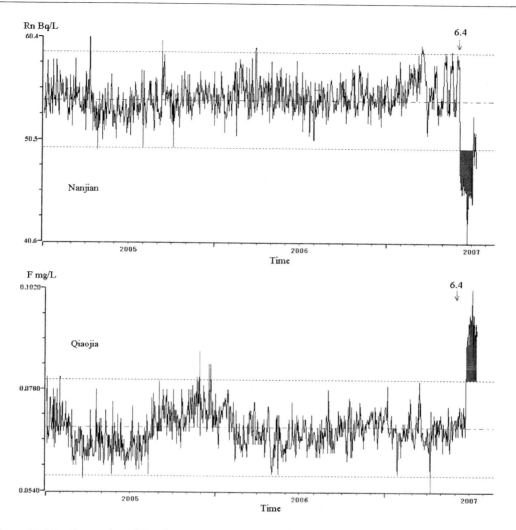

Figure 13: The time series of the observer components at nine observatories in the Chuan-Dian block, showing the geochemical anomalies related to the 3 June 2007 Puer M6.4 eathquake, the typed names in each diagram are corresponding to those in Table 1 and Figure 12.

The experimental data of rock-fluid interaction under the ambient condition demonstrate that (1) water-rock reaction can result in both measurable increase and decrease in ion concentrations in six hours or less, and (2) the extent of hydrochemical variation is controlled by the efficient surface of reaction (grain size), dissolution and secondary mineral precipitation, as well as chemical-types of rock and groundwater (Figure 14, 15). The results demonstrate water-rock reactions in brittle aquifer and aquitard may be an important genetic process of hydrochemical seismic precursors when the aquifer and aquitard are fractured under action of tectonic stress [115].

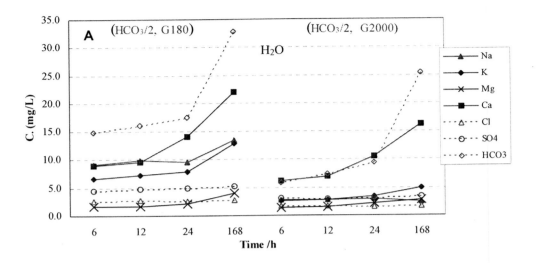

Figure 14: Ion concentrations of the experimental solutions for soaking granodiorite grains of 2000 and 180 μm in diameter with deioned water from 6 to 168 hours, showing the two-steps model: (1) water mixed with fluids of fluid inclusions in rock resulted in soluble Cl and SO_4 approached approximately equilibrium in short time (6 h or less); (2) dissolution predominantly controlled concentrations of other ions, resulting in the concentrations increased with increasing soaking time.

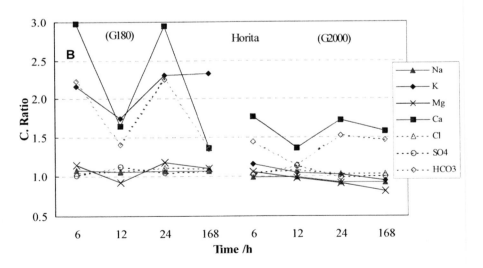

Figure 15: Variations of ion concentrations with time. The ordinate denotes corresponding ion concentration ratios of the experimental solutions over Horita geothermal water. The concentrations of Ca, K and HCO_3 increased 20% to 200% appear to be more sensitive for formation of micro-fracture.

A conceptual model (Figure 16), in which the role of fluids are emphasized, is proposed here based on the geological and geophysical data and the experiments of rocks at high pressure and high temperature in order to highlight mechanism of earthquake and formation of precursory anomalies in the continental region. Deep-earth fluids with high temperature carry plenty of energy into the lithosphere by migrating upwards and accumulate in some layers forming the zones of lower velocity and higher conductivity (the red horizontal parts in Figure 16). The fluids can escape by crypto-explosion from the accumulation when the tress that results from the fluids and heating in the accumulation exceeds the strength of host rocks. The upper crust is the most important boundary layer of the solid earth, in which the active faults affect the boundary conditions of the accumulation and migration of deep fluids related to seismic activity, especially medium and deep earthquakes. Rocks in the deep earth (>10 km) are not brittle, but are ductile and behave rheologically [116]. The fuci of most earthquakes in China's mainland are deeper than 15 km. Most faults, called as seismic faults, disappear in the detached zone at depth of about 11 km in north China. In this case, it is difficult to explain the generation of earthquakes by the elastic rebound theory of earthquake [117]. The model proposed here can better explain the mechanism of medium and deep earthquakes and some geochemical seismic precursors. This model is supported by many evidences:

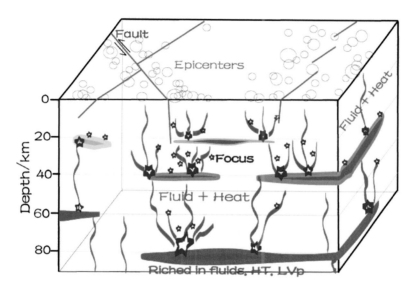

Figure 16: A model for generations of earthquake and fluid geochemical precursor, showing that deep-earth fluids carrying plenty of energy migrate upwards and accumulate in some layers in the lithosphere forming the zones of lower velocity and higher conductivity (the red horizontal parts). The fluids escape by crypto-explosion from the accumulation when the tress that results from the fluids and heating in the accumulation exceeds the strength of host rocks. The stars stand fro foci of earthquakes and its size for different magnitudes. The red lines on the top of the block indicate active faults which affect the boundary conditions of the fluid accumulation and act as the passage for fluids.

1) The zones with lower P-wave velocity in the crust and upper mantle are ubiquitous found in the intraplate seismic regions [110, 118, 119], nearby which the foci of great earthquakes with M>6.0 in China's continent are located. For example, the several great history earthquakes (1303 Hongdong, Shanxi M8.0, 1556 Huaxian, Shaanxi M8.0, 1668 Tancheng M8.5, 1679 Sanhe-Pinggu M8.0 and 1920 Haiyuan, Gusu M8.6 earthquakes) occurred in northern China and vicinity are correlated with the lower velocity zones [120]. The 4th February 1975 Haicheng M7.3 earthquake and the after shocks occurred nearby the top of the zone with lower-velocity and high-

conductivity (Figure 17)[1] [23]. The foci of three large earthquakes, 15th November 1976 Ninghe M6.9, 28th July1976 Tangshan M7.8 and Luanhe M7.1 earthquakes, also occurred nearby the lower velocity zones (Figure 18) [12, 121]. Similarly, the fuci of the October 1989 Datong-Yanggao earthquakes and 8 January 1998 Zhangbei M6.2 earthquake are close to the zones with lower velocity and high conductivity [122].

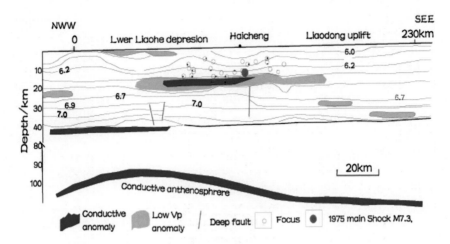

Figure 17: The profile of velocity structure across Haicheng, Liaonin Province, showing the spatial relationship between the zones of lower velocity and higher conductivity and the fuci of the 4th February 1975 Haicheng M7.3 earthquake and after shocks (modifies after Liu et al.).

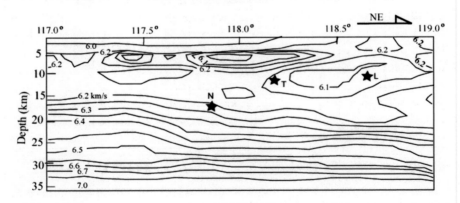

Figure 18: The profile of crustal structure across Tangshan, isolines of absolute wave velocity display the zones of lower velocity, nearby which three earthquakes occurred, stars marked N, T and L indicate fuci of Ninghe, Tangshan and Luanhe earthquakes, respectively [12, 121].

2) A plenty of fluids do exist in the deep earth, which have an important effect on a whole range of physical and chemical properties of the mantle including rheology, partial melting, diffusion, electrical conductivity, and seismic wave speeds and attenuation [123-126]. The fluids and heat in the earth are the main factors that

---

[1]Liu et al., (1995). The lithospheric structure, the physical and 3-D digital models for intraplate earthquake generation and prediction for potential seismic focus area. An unpublished report, in Chinese.

control geothermic resources, volcanic, seismic and tectonic activities. The deep earth fluids as the carrier of the heat migrate upwards, providing energy for geothermal field, magma generation, volcanic eruption and earthquake generation. Fault and seismic activities inversely favor migration of deep-earth fluids. Earthquakes frequently occur in the hydrothermal zones that overlap the active tectonic regions. The four global geothermal zones mostly overlap the global seismic zones [12]. Consequently, migration of fluids from the deep earth to the surface provides energy for earthquake generation and results in geochemical anomalies.

3) The crypto-explosion is the phenomena that were observed worldwide. The crypto-explosion granodiorite breccias contacted with intrusive granodiorite body are observed in Yanqing, Beijing. Many crypto-explosion breccias related to core deposits have been found worldwide [127, 128]. The fluids derived from the mantle are enriched in alkali components and volatiles. The expansion of volatiles usually results in crypto-explosion breccias during the magma migration. In addition, basalt with big mantle xenolith in Hannoba, Hebei Province, north China indicates the magma was derived from the upper mantle and migrated upwards at high speed. Therefore, it is evident that fluids in the deep earth can create the way for migration in which earthquake is generated.

4) The experiments of gabbro, anorthosite and pyroxenite at high temperature and high pressure indicate that fluids resulted from mineral dehydration can produce the zones of lower velocity and high conductivity in the lithosphere [129-132]. The recent mineral physics and seismological observations suggest that water significantly reduces seismic wave velocities through anelastic relaxation. The calculation for seismic wave velocities and attenuation in the upper mantle with a range of water contents indicates that the sharp velocity change around 60~80 km (the Gutenberg discontinuity) is attributed to a sharp change in water content due to partial melting [124]. The dehydration of serpentine at high temperature and high pressure resulted in decrease of the ultrasonic velocity and explosion in the experimental cell with loud sound, which is similar to occurrence of earthquake (Figure 19) [133].

5) The geochemical variations were observed after the nuclear explosion and the field test of explosion (Figure 20) [134]. Similarly, the seismic geochemical anomalies can result from the crypto-explosion caused by expansion of fluid derived from the deep earth. Such progress can be used to explain the formation of geochemical anomalies after great earthquake.

6) Gases derived from the crust and mantle usually exist in the seismic zones in China [11, 108, 135- 138], which act as the carrier for trace amount of gases. Precursory anomalies of gases, such as radon, mercury, helium, hydrogen, carbon dioxide, both in free gas and water, are usually characterized by "spike-like increase" with a precursory duration from several hours to several months before the great earthquakes. For instance, the multidisciplinary investigation in the Vogtland region, Germany, suggests that gas component of the observed fluid (99 vol. % CO_2) is of upper mantle/crustal origin and the seismically active faults act as the fluid transport pathway to the surface, suggesting an influence of the fluid system due to the processes of earthquake generation. The very rapid recovery of atmospheric radon concentration to the previous normal low level in October 1994 just after the earthquake seems to indicate a rapid emanation of radon atoms in rocks released to

ambient water in the crust due to activities of faults and fissures, as shown by large hourly radon concentrations observed just before the earthquake onset [139]. Such phenomena are ascribed to the process of terrestrial gas emanation, as proposed by King (1986) [30], Luo et al. (2002) [25] and Ondoh & Hayakawa (2006) [28]. It is impossible for radon and trace amount gasses to migrate by diffusion for a long distance in a short time because they have very low concentrations, and radon has a short half-life time (3.85 d). Trace amount gasses that disperse in minerals can diffuse or emit into underground water and free gas in fractures and pores of rocks under the action of tectonic stress related to the seismic events and then migrate upwards with fluids, in which water and carbon dioxide may act as the carrier. Therefore, the deep fluid carriers, such as H_2O, CO_2, CH_4, He, N_2, are necessary for ^{222}Rn and Hg to transport for a long distance in a short time, as reported by Martinelli and Ferrari (1991) [140].

7) The long-term seismic precursors of gasses characterized by continuous increase may be due to a gradual accumulation of fluids in the solid earth, resulting in gradual growth of seismic structure changes. The fault activities and wider fissures due to an enhanced tectonic strain before the earthquake will produce crushed rock fragments around the active faults and fissures which are inversely favorable for migration of fluids and changing the concentrations of chemical components.

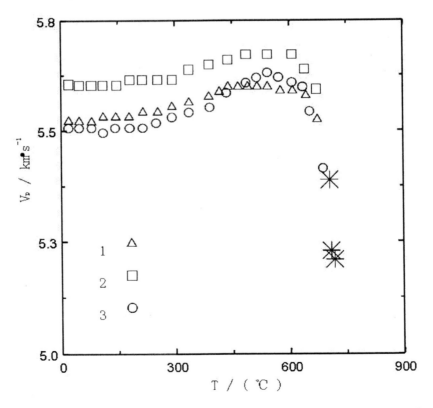

Figure 19: Plot of P-wave velocity versus temperature at 1, 2, and 3 GPa corresponding to the numbers in the legend, illustrating explosion (star) caused by dehydration of serpentine [133].

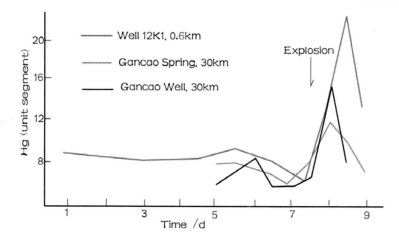

Figure 20: The concentration of mercury in the well and spring waters in northwestern China increased obviously after nuclear explosion. The well No.12K1 is at a distance 0.6 km to the nuclear explosion site, the Gancao Well and Gancao Spring at a distance of 30 km [134].

CONCLUSION

Many devastating earthquakes occurred in China, which left many deaths and caused a lot of economic loss. Earthquake prediction is considered as one of the most efficient approaches to mitigate seismic hazard. Predicting earthquake depends on understanding the genetic processes of earthquake and seismic precursors. Geochemical monitoring for earthquake is an efficient technique for earthquake prediction. The results indicate the geochemical anomalies related to great earthquakes do exist, and selection of suitable approaches for processing data is critical. However, it is still a puzzle to distinguish seismic precursors from the whole of geochemical anomalies.

The monitoring network has been improved in China. Plenty of novel geochemical data in China were obtained since the late 60's of last century. Correlation between the geochemical anomalies and great earthquakes occurred in China indicates that the temporal and spatial features of geochemical anomalies connected with great earthquakes vary dramatically. There are a few of successful examples for predicting earthquakes based on the gaseous and hydrochemical anomalies, but a lot of failure ones.

A model for generations of earthquake and seismic precursors of geochemistry is proposed. The migration of deep-earth fluids by emanation and crypto-explosion is one of important mechanism for generations of the geochemical anomalies and earthquakes. The hydrochemical anomalies related to earthquake can be attributed to tow processes of fluid mixture and water-rock interaction that are cause by crustal stress.

ACKNOWLEDGEMENT

The authors would like to thank Senior Engineers Yongxin Shao in Tianjing Bureau of Seismology, Suxin Zhang in Hebei Bureau of Seismology, Xuefang Fan in Shanxi Bureau of

Seismology, Zhijun Guan in Sichuan Bureau of Seismology, Xiaofeng Liu and Prof. Liming Yang in Gansu Bureau of Seismology for presenting the dear observation data. This work was supported by Earthquake Science Foundation (No.B07002), the Seismological Fund of Institute of Earthquake Science, CEA. The first author is grateful to the Institute of Geochemistry, CAS for supplying the nice conditions of live and work when he works in Guiyang as a visiting professor.

REFERENCES

[1] Barsukov, V.L., Varshal, G.M. & Zamokina, N.S. (1985a). Recent Results of Hydrogeochemical Studies for Earthquake Prediction in the USSR. *Pure Appl. Geophys.* 122, 143-156.

[2] Wakita, H., Nakamura, Y. & Sano, Y. (1988). Short-term and intermediate-term geochemical precursors. *Pure Appl. Geophys.* 125, 267–278.

[3] Thomas, D. (1988). Geochemical precursors to seismic activity. *Pure Appl. Geophys.* 126, 241–266.

[4] Toutain, J.P. & Baubron, J.C. (1999). Gas Geochemistry and seismotectonics: a review, *Tectonophysics* 304, 1–27.

[5] Biagi, F.P., Ermini, A. Kingsley, S.P., et al. (2000a), Groundwater ion content precursors of strong earthquakes in Kamchatka (Russia), *Pure Appl. Geophys.* 157, 1359-1377.

[6] Biagi, P.F., Ermini, A., Kingsley, S.P., et al. (2000b). Possible precursors in groundwater ions and gases content in Kamchatka (Russia). *Phys. Chem. Earth (A),* 25(3), 295-305.

[7] Kingsley, S.P., Biagi, P.F., Piccolo, R. et al. (2001). Hydrogeochemical precursors of strong earthquakes: a realistic possibility in Kamchatka, *Phys. Chem. Earth* 26, 769-774.

[8] Favara, R., Grassa, F., Inguaggiato, S., et al. (2001). Hydrogeochemistry and stable isotopes of thermal springs: earthquake-related chemical changes along Belice Fault (Western Sicily), *Applied Geochem.* 16, 1-17.

[9] Belin, B., Yalçin, T., Suner, F., et al. (2002). Earthquake-related chemical and radioactivity changes of thermal water in Kuzuluk-Adapazar, Turkey. *J. Environ. Radioact.* 63, 239–249.

[10] Du, J., Li, Z., Wang, C., et al. (2003). Seismological significance and features of fluid anomalies occurred in Hohhot City in 2003. *Earthquake* 24(1), 27-33. (in Chinese with English abstract)

[11] Du J., Cheng, W., Zhang, y., et al. (2006). Helium and Carbon Isotopic Compositions of Thermal Springs in Earthquake Zone of Sichuan, Southwestern China, *J. Asia Earth Sci.* 26, 533-539.

[12] Du, J., Zhang, Y. & Li, H. (2007). Advances in studies of thermal-fluid geochemistry and hydrothermal resource in China. In: Ueckermann, H.I. (Ed), *Geothermal Energy Research Trends*, New York: Nova Science Publishers, Inc., pp. 51-88.

[13] Okabe, S. (1956). Time variation of the atmospheric radon content near the ground surface with relation to some geophysical phenomena. *Mem. Coll. Sci. Univ. Kyoto* Ser. A. 28 (2), 99-115.

[14] Ghosh, D., Deb, A., Sengupta, R., et al. (2007). Pronounced soil-radon anomaly — precursor of recent earthquakes in India, *Radiation Measurements*, doi:10.1016/j.radmeas. 2006.12.008.

[15] Rikitake, T. (1976). *Earthquake Prediction*, Amsterdam: Elsevier, 1-357.

[16] Rikitake, T. (1980). Earthquake and its forecasting. *GeoJournal* 4(2), 145-152.

[17] Li, G., Jiang, F., Wang, J., et al. (1985). Preliminary Results of Seismogeochemical Research in China. *Pure Appl.Geophys.* 122, 218-230.

[18] Niazi, M. (1985). Regression analysis of reported earthquake precursors. I. Presentation of data. *Pure Appl. Geophys.* 122, 966-981.

[19]]Zhang, W., Wang, J. & X-M. E (1988a). *Principles and Methods of Hydrogeochemical Prediction for Earthquakes*, 1-272, Beijing Education Press, Beijing (in Chinese).

[20] King, C.Y., Basler, D., Presser, T.S., et al. (1994). In search of earthquake-related hydrologic and chemical changes along Hayward fault. *Appl. Geochem.* 9, 83–91.

[21] Léonardi, V., Kharatian, K., Igumnov, V., et al. (1999). Changes in water level, chemistry and helium emission induced by seismicity in confined aquifers in Armenia, *Earth Planet. Sci. 328*, 51-58.

[22] Du, J. & Kang, C. (2000a). A brief review on study of earthquake-caused change of underground fluid, *Earthquake 20(Sup.)*, 107-104 (in Chinese with English abstract).

[23] Du, J. & Kang, C. (2000b). Characteristics of earthquake precursors and its geological significance. *Earthquake* 20(3), 95-101 (in Chinese with English abstract).

[24] Wu, Z. & Cai, Y. (2001). Corresponding relation between ground hydrochemistry anomalies and earthquake at Chaozhou station, *South China J. Seism. 21(1)*, 38-42 (in Chinese with English abstract).

[25] Luo, Z., Guo, D., Zhang, T., et al. (2002). *The New Methods and Techniques of Earthquake Prediction in the Earthquake-Test Fields*, 1-259, Beijing: Seismological Press (in Chinese).

[26] Claesson, L., Skelton, A., Graham, C., et al. (2004), Hydrogeochemical changes before and after a major earthquake, *Geology 32(8)*, 641–644.

[27] King, C.Y., Zhang, W. & Zhang Z. (2006). Earthquake-induced Groundwater and Gas Changes. *Pure Appl. Geophys.* 163, 633–645.

[28] Ondoh, T. & Hayakawa, M. (2006). Synthetic study of precursory phenomena of the M7.2 Hyogo-ken Nanbu earthquake. *Phys. Chem. Earth* 31,378–388.

[29] Wang, C. (1990). *Seismic Observation Net of Underground Water Regime*. Beinjing: Seismological Press (in Chinese).

[30] King, C.Y. (1986). Gas geochemistry applied to earthquake prediction: An overview. *J. Geophys. Res.* 91, 12269–12281.

[31] Fleischer, R. L. (1981). Dislocation model for radon response to distant earthquakes. *Geophys. Res. Letters* 8, 477-480.

[32] Reimer, G.M. (1985).Prediction of central California earthquakes from soil-gas helium fluctuations. *Pure Appl.Geophys.*122, 369-375.

[33] Planinić, J., Radolić, V. & Lazanin, Ž. (2001). Temporal variations of radon in soil related to earthquakes. *Appl. Radiat. Isot.* 55, 267–272.

[34] Dobrovolsky, I.P., Gersherzon, N.I. & Gokhberg, M.B. (1989). Theory of electrokinetic effect occurring at the final state in the preparation of a tectonic earthquake. *Phys. Earth Planet Interiors* 57, 144–156.

[35] Igarashi, G., Saeki, S., Takahata, N., et al. (1995). Ground-water radon anomaly before the Kobe earthquake in Japan, *Sci.* 269, 60.

[36] Sharma, S.C. (1997). Thermal springs as gas monitoring sites for earthquake prediction. In: Virk, H.S. (Ed), *Proc. 3rd Int. Conf. on Rare Gas Geochemistry, Amritsar, India, 10-14 December 1995*, pp. 193-199.

[37] Bolognesi, L. (2000).Earthquake-induced variations in the composition of the water in the geothermal reservoir at Vulcano Island, Italy. *J. Volcan. Geotherm. Res.* 99,139–150.

[38] Quattrocchi, F., Pik, R., Pizzino, L., et al. (2000). Geochemical changes at the Bagni di Triponzo thermal spring during the Umbria-Marche 1997–1998 seismic sequence. *J. Seismol.* 4, 567–587.

[39] Zmazek, B., Italiano, F., Živčić, M., et al. (2002). Geochemical monitoring of thermal waters in Slovenia: relationships to seismic activity. *Applied Radiation and Isotopes* 57, 919–930.

[40] Seewald, J., Cruse, A. & Saccocia, P. (2003).Aqueous volatiles in hydrothermal fluids from the Main Endeavour Field, northern Juan de Fuca Ridge: temporal variability following earthquake activity. *Earth Planet. Sci. Let.* 216, 575-590.

[41] Planinić, J., Radolić, V. & Vuković, B. (2004). Radon as an earthquake precursor. *Nucl. Instrum. Method Phys. Res. A* 530, 568–574.

[42] Erees, F.S., Aytas, S., Sac, M.M., et al. (2007). Radon concentrations in thermal waters related to seismic events along faults in the Denizli Basin,Western Turkey. *Radiat. Meas.* 42, 80 – 86.

[43] Shapiro, M.H., Rice, A., Mendenhall, M.H., Melivin, J.D., Tombrello, T.A. (1985). Recognition of environmentally caused variations in radon time series. *Pure Appl.Geophys.* 122, 309-326.

[44] Segovia, N., Cruz-Reyna, S.D.L., Mena, M., et al. (1989). Radon in soil anomaly observed at Los Azufres Geothermal Field, Michoacan: a possible precursor of the 1985 Mexico earthquake (Ms = 8.1). *Natural Hazards* 1, 319-329.

[45] King, C.Y., King, B.S., Evans, W.C., et al. (1996). Spatial radon anomalies on active faults in California. *Appl. Geochem.* 11, 497–510.

[46] Eichhubl, P. & Boles, J.R. (2000). Focused fluid flow along faults in the Monterey Formation, coastal California. *GSA Bulletin* 112(11), 1667–1679.

[47] Pili, É., Poitrasson, F. & Gratier, J.P. (2002).Carbon–oxygen isotope and trace element constraints on how fluids percolate faulted limestones from the San Andreas Fault system: partitioning of fluid sources and pathways. *Chem. Geol. 190*, 231– 250.

[48] Pizzino, L., Burrato, P., Quattrocchi, F., et al. (2004).Geochemical signatures of large active faults: The example of the 5 February 1783, Calabrian earthquake (southern Italy). *J. Seismol.* 8: 363–380.

[49] Yang, T.F., Walia, V., Chyi, L.L., et al. (2005). Variations of soil radon and thoron concentrations in a fault zone and prospective earthquakes in SW Taiwan. *Radiat. Meas.* 40, 496 – 502.

[50] Dubrovskiy, V.A. & Sergeev, V.N. (2006). Short- and medium-term earthquake precursors as evidence of the sliding instability along faults. *Phys. Solid Earth* 42(10), 802–808.

[51] Aki, K. (1981). Probabilistic synthesis of precursory phenomena, In: Simpson, D.W., Richards, P.G. (Ed). *Earthguake Prediction*, Washington D.C.: AGU 566-571.

[52] Barsukov, V.L, Serebrennikov, V.S., Beleyaev, A.A., et al. (1985b). Some Experience in Unraveling Geochemical Earthquake Precursors. *Pure Appl. Geophys.* 122, 157–163.

[53] Tarcotte, D.L. (1992). *Fractals and Chaos in Geology and Geophysics*, Cambridge: Combridge Univ. Press.

[54] Fu, H., Chen, L., Luo, P., et al. (1997) Medium-short- impending prediction and precursory anomalous characteristics on earthquake of M=7.3 on July12 of 1995 in border area between China Yunnan's Menlian and Burma: (3) Medium-short-impending anomaly characteristics for all fixed observatory items. *J. Seism. Res.* 20(14), 345-356 (in Chinese with English abstract).

[55] Cuomo, V., Di Bello, G., Lapenna, V., et al. (2000). Robust statistical methods to discriminate extreme events in geoelectrical precursory signals: implications with earthquake prediction. *Nat. Hazard* 21, 247–261.

[56] Biagi, P.F., Ermini, A., Kingsley, S.P., et al. (2001). Difficulties with interpreting changes in groundwater gas content as earthquake precursors in Kamchatka. *Russ. J. Seismol.* 5, 487–497.

[57] Negarestani, A., Setayeshi, S., Ghannadi-Maragheh, M., et al. (2001). Layered neural networks based analysis of radon concentration and environmental parameters in earthquake prediction. *J. Environ. Radioact.* 62, 225–233.

[58] Toutain, J.P., Munoz, M., Pinaud, J.L., et al. (2006). Modelling the mixing function to constrain coseismic hydrochemical effects: An example from the French Pyrénées. *Pure Appl. Geophys.* 163, 723–744.

[59] Ciotoli, G., Lombardi, S. & Annunziatellis, A. (2007). Geostatistical analysis of soil gas data in a high seismic intermontane basin: Fucino Plain, central Italy. *J Geophys Res*, 112, B05407, doi: 10.1029/2005JB004044.

[60] Hartmann, J. & Levy, J.K. (2005). Hydrogeological and Gasgeochemical Earthquake Precursors – A Review for Application. *Natural Hazards*, 34(3), 279-304.Du, J., Yuwen, X., Li, S., et al. (1998a). The geochemical characteristics of escaped radon from the Babaoshan fault zone and its earthquake reflecting effect. *Earthquake* 18(2), 155-162 (in Chinese with English abstract).

[62] Zmazek, B., Živčić, M., Todorovski, L., et al. (2005). Radon in soil gas: How to identify anomalies caused by earthquakes. *App. Geochem.* 20, 1106–1119.

[63] Wang, X., Shi, S. & Ding, X. (1999). Bayes discriminant analysis and its application to short and imminent earthquake prediction. *Earthquake* 19(1), 33-40 (in Chinese with English abstract).

[64] Slemmons, D.B. (1982). Determination of design earthquake magnitudes for Microzonation. *Proceedings of 3rd International Earthquake Microzonation Conference, Seattle* 1, 119-130.

[65] Rikitake, T. (1987). Earthquake precursors in Japan: precursor time and detectability. *Tectonophysics,* 136, 265-282.

[66] Dobrovolsky, I.P., Zubkov, S.I. & Miachkin, V.I. (1979). Estimation of the size of earthquake preparation zones. *Pure Appl. Geophys.* 117, 1025–1044.

[67] Sultankhodzhayev, A.N., Latipov, S.U., Zakirov, T.Z., et al. (1980). Dependence of Hydrogeoseismological Anomalies on the Energy and Epicentral Distance of Earthquakes. *Dokl. AN Uzb. SSR* (5), 57-59.

[68] Wang, C., Tao, S., Zhang, S., et al. (1996). The Characteristics of hydrochemical precursory field for great earthquakes. *Northwestern Seismolo. J.* 18(1):1-7 (in Chinese with English abstract).

[69] Wang, J., Zheng, Y. & Zhang, S. (1998). Analysis of the hudrochemical precursory anomalies and discussion of the related issues of Zgangbei M6.2 earthquake. North China Earthquake Sci. 16(3), 46-52 (in Chinese with English abstract).

[70] Lin, H., Wang, S. & Fu, H. (2006). Comprehensive analysis of underground fluid precursors at Bangnazhang Hotspring, Longlin, Yunnan. *J. Seism. Res.* 29(1), 35-38.

[71] Wang, D. (2004). The dynamic abnormal variation of underground fluid in the North Tianshan Mountain before and after earthquakes with more than Ms7 in Xinjiang and its adjacent areas. *Inland Earthquake* 18(1), 45-55 (in Chinese with English abstract).

[72] Wang, R., Woith, H., Milkeveit, C., et al. (2004). Modelling of hydrogeochemical anomalies induced by distant earthquakes *Geophys. J. Int.* 15, 7717-726.

[73] Huang, F., Jian, C., Tang, Y., et al. (2004). Response changes of some wells in the mainland subsurface fluid monitoring network of China, due to the September 21, 1999, Ms7.6 Chi-Chi Earthquake. *Tectonophysics* 390, 217– 234.

[74] Ye, X., Yang, M. & Jia, Q. (2004). The features of hydrochemical precursor in Guangdong and Fujian Provinces, *Seism. Geomagn. Obser. Res.* 25(6), 17-23 (in Chinese with English abstract).

[75] Teng, T.L. (1980). Some recent studies on groundwater radon content as an earthquake precursor, *J. Geophy. Res. 85 (B6)*, 3089.

[76] Ma, Z., Fu, Z., Zhang, Y., et al. (1982). *Nine Great Earthquakes in China (1966–76)*. Beijing: Seismological Press, 1-216.

[77] Mei S., Feng, D., Zhang G., et al. (1993). *Introduction to Earthquake Prediction in China*. Beijing: Seismological Press, 1-246 (in Chinese).

[78] Ye, X. (1998). Analysis of the subsurface fluid precursor anomaliy before Taiwan Strait earthquake of Ms7.3. *South China J. Seismol.* 18(3), 35-40 (in Chinese with English abstract).

[79] Song, S.R., Chen, Y.L., Liu, C.M., et al. (2005). Hydrochemical Changes in Spring Waters in Taiwan: Implications Evaluating Sites for Earthquake Precursory Monitoring. *TAO*, 16(4), 745-762.

[80] Fu, H., Liu, L. & Zgang, X. (2005). Research on seismicity of group earthquakes and judge indexes in Chuan-Dian Arae. *Earthquake* 25(1), 8-14 (in Chinese with English abstract).

[81] Zhang, Z., Luo, L. & Li, H. (1988b). *China Earthquake Cases (1966-1975)*, Beijing: Seismological Press (in Chinese).

[82] Zhang, Z., Luo, L. & Li, H. (1990). *China Earthquake Cases (1966-1975)*, Beijing: Seismological Press (in Chinese).

[83] Zhang, Z., Zheng D. & Xu, J. (1999). *China Earthquake Cases (1966-1975)*, Beijing: Seismological Press (in Chinese).

[84] Zhang, Z., Zheng, D. & Xu, J. (2000). *China Earthquake Cases (1976-1991)*, Beijing: Seismological Press (in Chinese).

[85] Chen, Q., Zheng, D. & Che, S. (2002a). *China Earthquake Cases (1992-1994),* Beijing: Seismological Press (in Chinese).

[86] Chen, Q., Zheng D. & Che, S. (2002b). *China Earthquake Cases (1995-1996),* Beijing: Seismological Press (in Chinese).

[87] Kayal, J.R. (1991). Earthquake prediction in northeast India--A review. *Pure Appl. Geophys* .136(2/3), 297-313.

[88] Wyss, M. (1997a). Cannot earthquakes be predicted?. *Science* 278, 487–488.

[89] Geller, R.J., Jackson, D.D., Kagan, Y.Y. et al. (1997). Response to M. Wyss: cannot

[90] Sykes, L.R., Shaw, B. & Scholz, C.H. (1999). Rethinking Earthquake Prediction. *Pure Appl. Geophys.* 155, 207–232.

[91] Roeloffs, E.A. (1988). Hydrologic precursors to earthquakes: a review. *Pure Appl. Geophys.* 126, 177-209.

[92] Roeloffs, E.A. (1996). Poroelastic techniques in the study of earthquake-related hydrologic phenomena. *Adv. Geophys.* 37, 135–195.

[93] Igarashi, G. & Wakita, H. (1995). Geochemical and hydrological observations for earthquake prediction in Japan. *J. Phys. Earth* 43, 585–598.

[94] Wyss, M. (1997b). Second round of evaluations of proposed earthquake precursors, *Pure App. Geophys.* 149, 3–16.

[95] Lin, Y. & Zhai, S. (1993). The method quickly measuring CO_2 in faul zone and its application in earthquake study. *Earthquake* 6, 65-67 (in Chinese with English abstract).

[96] Wu, Z. & Zhao, Z. (1995). The earthquake precursors characteristics of near source field for groundwater chemistry in western Yunnan area, *Northwest Seism. J.* 17(3), 33-42 (in Chinese with English abstract).

[97] Geng, J., Ma, Z., & Li, J. (1998). The discussion on characteristics and mechanism of seismic anomaly of many hydrogeochemical elements in Shiliquan Well before Cangshan M_S5.2 earthquake, *North China Earthquake Sci.* 16(1), 36-40 (in Chinese with English abstract).

[98] Tong, Z., Wang, Y., Liu, J., et al. (2000). Abnormal features of underground fluid in Panjin prior to Xiuyan-Haicheng M_S 5.6 earthquake, *Seism. Res. Northeast China* 16(2), 71-78 (in Chinese with English abstract).

[99] Yan, R., Huang, F. & Gu, J. (2004). Spatial-temporal characteristics of precursory anomaly of underground fluid before Ms7.0 strong earthquakes in China's continent. *Earthquake* 24(1), 26-31 (in Chinese with English abstract).

[100] Luo, P. (1991). Radon anomaly in the Nov. 6, 1988 Lanchang-Gengma earthquake. *J. Seismol Res.* 14(1), 41-49.

[101] Du, J. & Liu, C. (2003). Isotopic-geochemical application to earthquake prediction. *Earthquake* 23(2), 99-107 (in Chinese with English abstract).

[102] Yang, J. (1997). Hydrochemistry anomaly prior to Wuding M6.5 earthquake. *South China J. Seismol.* 17(1), 50-57 (in Chinese with English abstract).

[103] Yang, J. (1999). Analysis of precursor anomaly characteristics of mercury in groundwater for far and near shocks in Yunnan, *South China J. Seism.* 21(1), 64-70 (in Chinese with English abstract).

[104] Lin, Y., Wang, J. & Gao, S. (1998). A new measurement method for CO_2 in fault gas and prediction of Zhangbei-Shangyi M6.2 earthquake. *Earthquake* 18(4), 353-357 (in Chinese with English abstract).

[105] Zhang, L. & Fu, H. (2000). Study on the relationship between M≥6.0 earthquakes and hydrochemistry precursory anomalies in Yunnan region. *Earthquake Res. Plateau* 12(1), 42-49 (in Chinese with English abstract).

[106] Song, S.R., Ku, W.Y., Ceng, Y.L., et al. (2006). Hydrogeochemical anomalies in the springs of the Chiayi Area in West-central Taiwan as possible precursors to earthquakes. *Pure Appl. Geophys.* 163, 675–691.

[107] Jiang, C., Zhou, R. & Yao, X. (1998). Fault structure of Tengchong volcano. *J. Seismol. Res.* 21 (4), 330–336 (in Chinese with English abstract).

[108] Du, J., Liu, C., Fu, B., et al. (2005). Variations of geothermometry and chemical-isotopic compositions of hot spring fluids in the Rehai geothermal field, Southwest China. *J. Volcano. Geotherm. Res.*, 142(3-4): 243-261.

[109] Yan, D., Zhou, Me., Wang, C.Y., et al. (2006). Structural and geochronological constraints on the tectonic evolution of the Dulong-Song Chay tectonic dome in Yunnan Province, SW China. *J. Asian Earth Sci.*, 28, 332–353.

[110] Huang, J., Zhao, D. & Zheng, S. (2002). Lithospheric structure and its relationship to seismic and volcanic activity in southwest China. *J. Geophys. Res.* 107(B10): 2255.

[111] Van der Hoven, S.J., Solomon, D.K. & Moline, G.R. (2005). Natural spatial and temporal variations in groundwater chemistry in fractured, sedimentary rocks: scale and implications for solute transport, *Applied Geochem. 20(5)*, 861-873.

[112] Koretsky, C. (2000). The significance of surface complexation reactions in hydrologic systems: a geochemist's perspective, *J. Hydrol. 230*, 127–171.

[113] Yu, J., Xu, F., Liu, W., et al. (2000). Experimental study on hydrodynamic model on seismic hydrochemical anomaly migration, *Earthquake* 21(1), 91-95 (in Chinese with English abstract).

[114] Toutain, J.P., Munoz, F., Poitrasson, F., et al. (1997). Springwater chloride ion anomaly prior to a M_L=5.2 Pyrenean earthquake, *Earth Planet. Sci. Let. 149*, 113-119.

[115] Du, J., Amita, K., Ohsawa, S., et al. (2008). Experimental evidence on formation of imminent and short-term hydrochemical precursors for earthquake. Geochemistry, Geophysics, Geosystems, (accepted).

[116] Yang, H. & Bai W. (2000). Progress of experimental study on rheology of lithosphere. *Progress Geophys.,* 15(2), 80-89.

[117] Reid, H.F. (1910). The mechanics of the earthquake. In: Lawson, A.C.(Ed) Report of the State Investigation Commission, Carnegie Institution of Washington, 2, 16-28.

[118] Fu, Z. (2000). The large-scale distribution of great shallow intraplate earthquakes in mainland China and adjacent areas and seismic coupling along plate margins. *J. Asian Earth Sci., 18*:41-46.

[119] Huang, J. & Zhao, D. (2004). Crustal heterogeneity and seismotectonics of the region around Beijing, China. *Tectonophysics,* 385:159–180.

[120] Zhang, X., Zhang, X.K., Liu, M., et al. (2003). Deep structural characteristics and seismo-gegesis of the M=8.0 earthquake. *Acta Seism. Sinica* 25(2), 136-142 (in Chinese with English abstract).

[121] Qi, C., Zhao, D., Chen, Y., et al.(2006). 3-D P and S wave velocity structures and their relationship to strong earthquakes in the Chinese capital region. *Chinese J. Geophy., 49(3)*: 805-815 (in Chinese with English abstract).

[122] Lai, X., Zhang, X. & Sun, Y. (2006). The complexity feature of crust—mantle boundary in Zhangbei seismic region and its tectonic implication. *Acta Seism. Sinica* 19(3), 243-250.

[123] Bell, D.R., Rossman, G.R. (1992). Water in Earth's mantle: the role of nominally anhydrous minerals. *Science* 255, 1391–1397.

[124] Karato, S. & Jung, H. (1998). Water, partial melting and the origin of the seismic low velocity zone in the upper mantle. *Earth Planet Sci Lett* 157,193–207.

[125] Grant, K., Ingrin, J., Lorand, J.P., et al. (2007). Water partitioning between mantle minerals from peridotite xenoliths. *Contrib. Mineral. Petrol.* 154, 15–34.

[126] Withers, A.C. & Hirschmann, M.M. (2007). H_2O storage capacity of $MgSiO_3$ clinoenstatite at 8–13 GPa, 1,100–1,400. *Contrib. Mineral. Petrol.* DO: 10.1007/s00410-007-0215-7.

[127] Sillltoe, R.H. (1985). Ore-related breccias in volcanoplutonic Arcs. *Econ. Geol.* 80(6), 1467-1514.

[128] Zhang, Z. (1991). General features and genetic mechanism of crypto-explosive breccias. *Geol. Sci. Tech. Inform.* 10(4), 1-5.

[129] Bai, L., Du, J., Liu, W., et al. (2002). The experimental studies on electrical conductivities and P-wave velocities of anorthosite at high pressure and high temperature. *Acta Seismol. Sinica* 15(6), 667-676.

[130] Bai, L, Du, J, Liu, W., et al. (2003). P-wave velocities and conductivity of gabbro at high pressures and high temperatures. *Sci. China (D)*, 46(9), 895-908.

[131] Liu, W, Du, J., Bai, L., et al. (2002). Compressional elastic wave velocities of serpentinized olivine-bearing pyroxenite up to 960□ at 1.0 GPa. *J. Phys. Conden. Matt.* 14, 11355-11358.

[132] Liu, W., Du, J., Ba, L., et al. (2003). Ultrasonic *P* wave velocity and attenuation in pyroxenite under 3.0GPa and up to 1170□. *Chinese Phys. Let.* 20(1), 164~166.

[133] Xie, H., Zhou, W., Li, Y., et al. (2000).The elastic characteristics of serpentinite dehydration at high temperature-high pressure and its significance. *Chinese J. Geophys.*, 43(6), 806-811.

[134] Zhang, G., Fu, Z. & Zhou X. (2001). *Introduction to Earthquake Prediction*, Beijing: Science Press, 1-1227.

[135] Du, J. (1994). Helium isotope evidence of mantle degassing in rift valley, Eastern China. *Chinese Sci. Bull.*, 39 (12), 1021-1024.

[136] Du, J., Xu, Y. & Sun, M. (1998b). $^3He/^4He$ and heat flow in oil and gas-bearing basins, China's mainland. *Chinese J. Geophys.*, 14 (2), 239-247.

[137] Shangguan, Z. (1995). Study of fluids and their stable isotope compositions from volcanic areas. In: Liu, R. (Ed), *Human Environment and Volcanic Action*. Seismological Press, Beijing, pp. 98–100 (in Chinese).

[138] Shangguan, Z., Bai, C. & Sun, M. (2000). Recent characteristics degassing from mantle-origin magma in Rehai of Tengchong county. *Sci. China, Ser. D: Earth Sci.* 30 (4), 407–414.

[139] Heinicke, J. & Koch, U. (2000).Slug flow—A Possible explanation for hydrogeochemical earthquake precursors at Bad Brambach, Germany. *Pure appl. geophys.* 157, 1621–1641.

[140] Martinelli, G. & Ferrari, G. (1991). Earthquake forerunners in a selected area of Northern Italy: recent developments in automatic geochemical monitoring. In: M. Wyss (Ed), Earthquake Prediction. *Tectonophysics,* 193: 397-410.

In: Geochemistry Research Advances
Editor: Ólafur Stefánsson, pp. 93-117

ISBN 978-1-60456-215-6
© 2008 Nova Science Publishers, Inc

Chapter 4

STRUCTURAL CHARACTERIZATION OF KEROGEN BY RUTHENIUM TETROXIDE OXIDATION

Veljko Dragojlovic
Wilkes Honors College of Florida Atlantic University, 5353 Parkside Drive, Jupiter, FL 33458

ABSTRACT

Kerogen is macromolecular sedimentary organic substance found in oil sands and oil shales. Kerogens of various types have been extensively studied by oxidative degradation methods. Main difficulties in reconstruction of the kerogen macromolecular structure, based on the products of an oxidative degradation, result from lack of information about the original structural elements. Over the past twenty years, ruthenium tetroxide oxidation, in combination with other analytical methods, has been extensively used in structural elucidation and has provided considerable insight into structure and origin of various kerogens. A somewhat unique feature of ruthenium tetroxide is that, although it is highly reactive, it is also a selective oxidizing agent. It is more selective compared to other common oxidants such as permanganate, chromate or ozone. Ruthenium tetroxide oxidizes alcohols, aldehydes, alkenes and alkynes to the corresponding carboxylic acids. Compared to the chromate or ozone oxidation, yields of ruthenium tetroxide oxidation are considerably better, which allows for a greater confidence in interpreting the results. While permanganate oxidation proceeds in similar yields, ruthenium tetroxide oxidation is considerably faster and simpler method. Furthermore, in some cases its chemoselectivity differs from that of permanganate and one can obtain additional information by combining the results of the two methods. A number of research groups used ruthenium tetroxide degradation to study composition of the aliphatic portion of the sedimentary organic matter. Nevertheless, in some cases, it is possible to obtain high yields of aromatic products. One of its most useful and unique features of ruthenium tetroxide is that it oxidizes ethers to the corresponding esters, thus allowing differentiation of ether moieties from esters. It has led to an extensive study of extant and fossil algaenans, which culminated into development of a method for their chemical fingerprinting.

1. INTRODUCTION

1.1. Kerogen

Kerogen is the insoluble macromolecular organic material dispersed in sedimentary rocks. It is the main form of organic sediments present in Earth's crust. Two detailed reviews of kerogen are available [1,2]. Knowledge of its structure is important to understand the nature and origin of sedimentary organic matter and to understand the process of oil genesis. Structural elucidation of kerogen is difficult because of its complexity and insolubility. While kerogen forms large macromolecules, it does not have a polymeric structure. Due to its complexity, one cannot rely on a single analytical technique and it is necessary to apply a multi-technique approach. Insolubility of kerogen means than a number of analytical methods and techniques that are routinely used in structural elucidation of organic material cannot be applied. A number of analytical techniques are based on degradation of kerogen macromolecule into soluble fragments, which are then analyzed by conventional means.

1.2. Methods for Elucidation of Structure of Kerogen

A variety of degradation methods have been applied to analysis of kerogens including reduction, hydrolysis, pyrolysis and various oxidation methods [2]. Among the oxidative degradations, alkaline permanganate oxidation has been one of the most widely used [3,4]. Modes of alkaline permanganate oxidation of different functional groups are well known. In an oxidative degradation of kerogen by this reagent, a variety of carboxylic acids are isolated in high yields. They are usually converted to the corresponding methyl esters and analyzed by GC-MS. Main difficulties in the reconstruction of the kerogen macromolecular structure, based on the products of permanganate oxidative degradation, result from a lack of selectivity of the reagent. Ruthenium tetroxide is a more selective oxidizing agent compared to permanganate and is likely to provide more reliable information about the affected centers – the original functional groups. Ruthenium tetroxide also has an advantage that it is used under mild conditions (neutral pH and room temperature), which minimize formation of artifacts. Among the oxidizing agents, ruthenium tetroxide is somewhat unique in that it is used in a mixture of water and organic solvents. Although a mixture of solvents is needed to dissolve the co-oxidant (water-soluble) and organic substrate, it has an added advantage in that presence of both water and organic solvents makes degradation more efficient as it solubilizes the diverse oxidation products better than a single solvent.

2. A BRIEF REVIEW OF RUTHENIUM TETROXIDE OXIDATIONS OF ORGANIC COMPOUNDS

Ruthenium tetroxide can be prepared by mixing a solution, or a suspension, of ruthenium salt in water with an oxidizing agent, such as sodium metaperiodate, followed by an extraction of ruthenium tetroxide with an organic solvent. Due to cost, as well as safety considerations, stoichiometric ruthenium tetroxide oxidations have been replaced by catalytic processes in which a catalytic amount of a ruthenium compound (most commonly hydrated $RuCl_3$ or RuO_2) is used together with a suitable stoichiometric co-oxidant. Role of the co-oxidant is to re-oxidize lower valent ruthenium species to ruthenium tetroxide.

Ruthenium tetroxide was introduced by Djerassi and Engle as an oxidizing agent for organic compounds [5]. Two reviews of ruthenium tetroxide oxidations of organic compounds are available in the older literature [6,7]. Ruthenium tetroxide is suitable for reactions which require a vigorous oxidizing agent under mild reaction conditions and, in this respect, it is superior to most other oxidants. As a powerful oxidizing agent, ruthenium tetroxide cannot be used without a solvent. In the past, ruthenium tetroxide had been known as a capricious oxidant, which would often produce very low yields when used in a stoichiometric amount. At the same time, there were difficulties in reproducing the catalytic reactions as they would sometimes stop due to deactivation of the catalyst. A milestone in development of an inexpensive and reliable catalytic ruthenium tetroxide oxidation has been introduction of acetonitrile as a co-solvent [8]. Ruthenium tetroxide is now firmly established as an effective and dependable organic oxidant. Since the introduction of acetonitrile as a co-solvent, ruthenium tetroxide oxidations of organic compounds have not been reviewed.

Catalytic ruthenium tetroxide oxidations are done in a biphasic solvent system consisting of water, acetonitrile and an organic solvent. Commonly used organic solvents are chlorinated solvents such as carbon tetrachloride, chloroform and dichloromethane. Recently, use of oxygenated solvents such as ethyl acetate and dimethyl carbonate has become more common. If one wants to do the reaction in a single phase, acetone can be used as the organic solvent. Some of the common organic solvents (ethers, aromatics) are oxidized by ruthenium tetroxide and cannot be used.

2.1. Oxidation of Alkanes

Ruthenium tetroxide is capable of oxidizing alkanes. Oxidation is relatively slow and either takes days at room temperature, or several hours under a reflux, to complete. The order of reactivity is methyne > methylene >> methyl. Yields are usually low and selectivity is poor, however, the reaction is of preparative value in certain cases (Figure 1) [9-11].

Cyclopropane ring activates the neighboring methylene group towards ruthenium tetroxide oxidation (Figure 2) [12,13].

2.2. Oxidation of Alkenes

Ruthenium tetroxide cleaves carbon-carbon double bonds to give the corresponding carbonyl compounds. Oxidation of a tetrasubstituted double bond results in the formation of a pair of ketones. When a double bond has less than four substituents, at least one of the products is a carboxylic acid (Figure 3) [8,14].

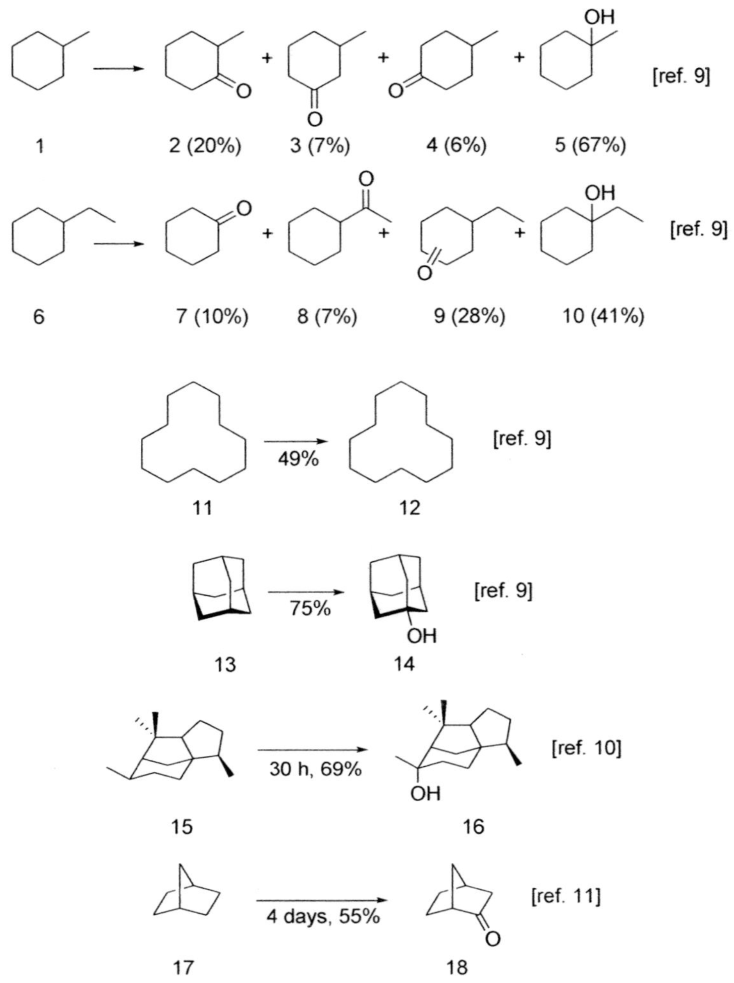

Figure 1.

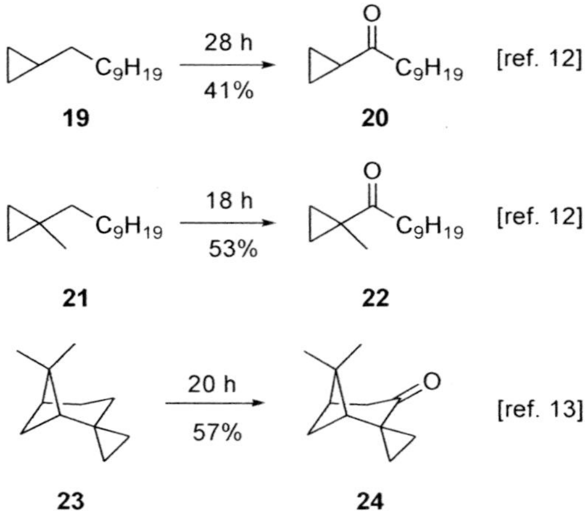

Figure 2.

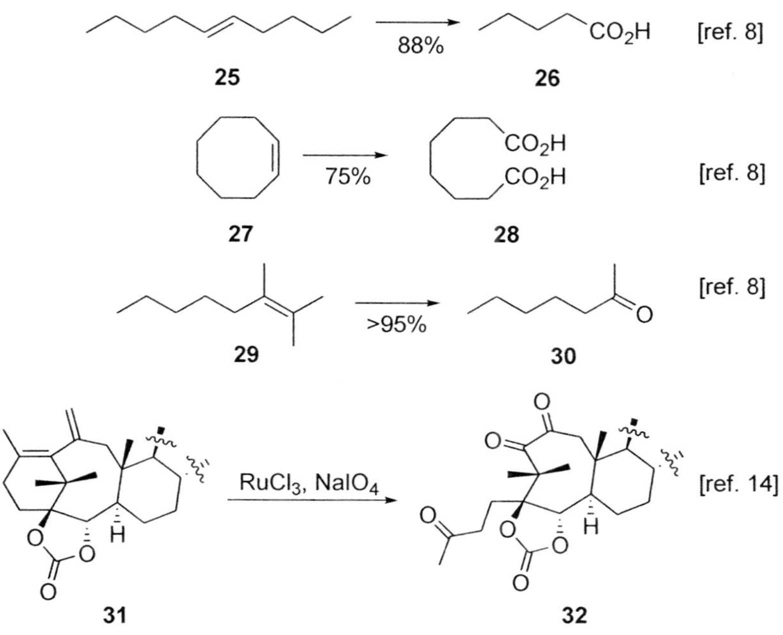

Figure 3.

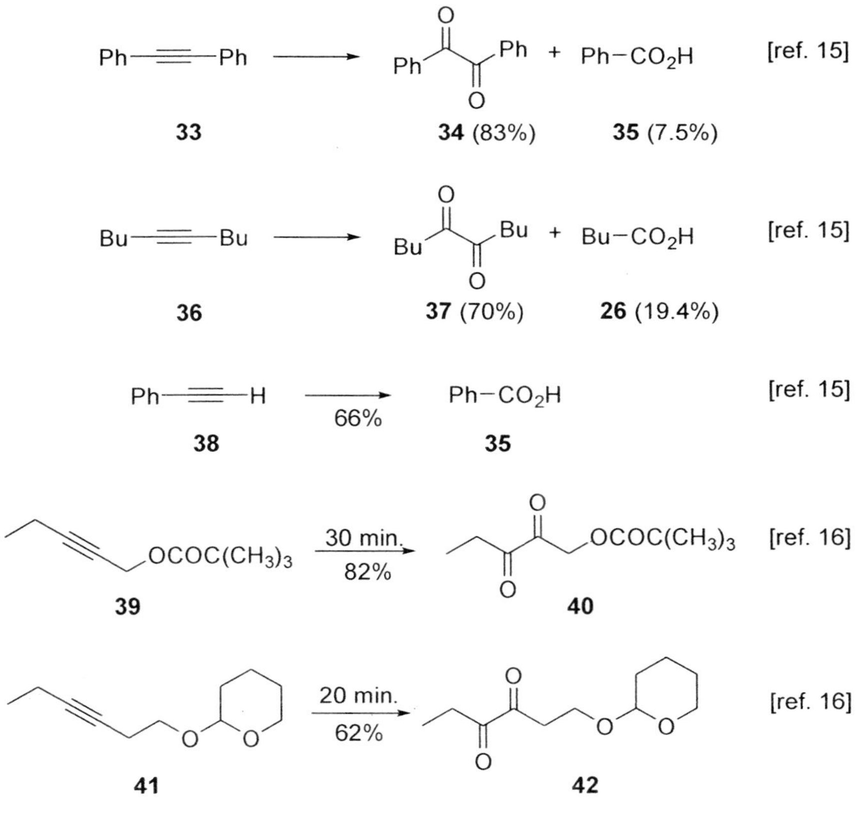

Figure 4.

2.3. Oxidation of Alkynes

Internal alkynes are oxidized by ruthenium tetroxide to the corresponding α-diketones. Cleavage of the diketone to the carboxylic acid is the main side reaction (Figure 4). Terminal alkynes are oxidized to the corresponding carboxylic acids [15].

Alkynes can be oxidized in the presence of esters as well as some ethers [16]. However, although alkynes are more reactive than ethers, the difference in reactivities is not very large and often both functionalities are oxidized. Some chemoselectivity can be achieved by keeping the reaction times short. Even then, yields of the selectively oxidized products are frequently low [16]. Alkenes, alcohols and arenes (not conjugated to the alkyne) are oxidized at approximately the same rate as alkynes [17].

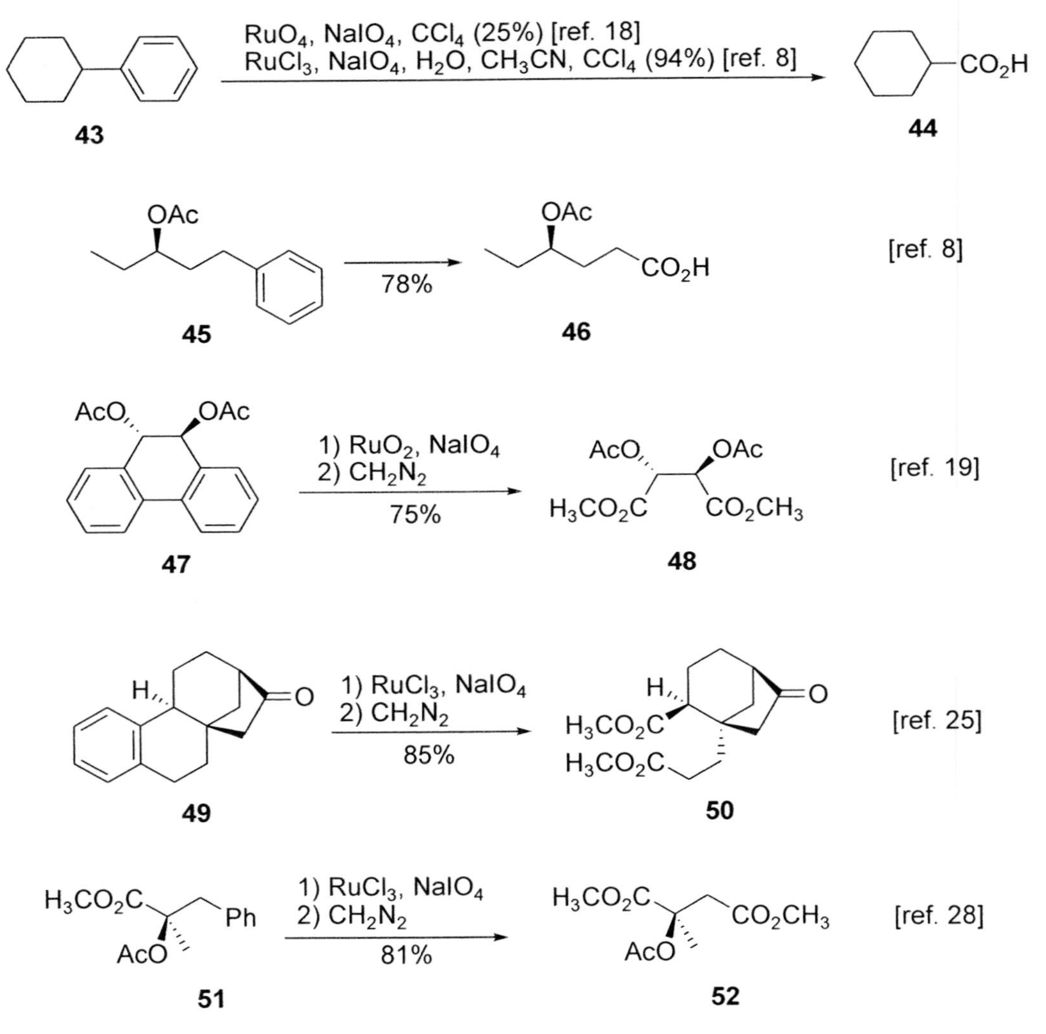

Figure 5.

2.4. Oxidation of Arenes

Ruthenium tetroxide oxidizes alcohols, aldehydes, alkenes, alkynes and some aromatic substituents to the corresponding carboxylic acids in a manner similar to that of permanganate. However, in oxidation of aromatic compounds its chemoselectivity differs from that of permanganate. Thus, oxidation of substituted benzenoic hydrocarbons with permanganate gives aromatic acids, while ruthenium tetroxide produces aliphatic acids with one more carbon atom (Figure 5). An aromatic ring is oxidized to the corresponding carboxylic acid with the ring itself being degraded to carbon dioxide and water. In that respect, mode of oxidation of aromatic rings by ruthenium tetroxide is similar to that of ozonolysis with an advantage that it gives considerably higher yields. If the reaction is carried out without acetonitrile only poor yields are obtained [18]. The Sharpless procedure for ruthenium tetroxide oxidation of aromatic rings provides excellent yields [8,19].

However, in some cases, as in the oxidation of compounds **53**, oxidation of phenyl ring has been reported to be sluggish and yields are modest [20].

$$\text{53} \xrightarrow[\substack{\text{1) HCl, THF} \\ \text{2) RuO}_2\text{, NaIO}_4 \\ \text{3) CH}_2\text{N}_2 \\ 56\%}]{} \text{54}$$

Reactivity of ruthenium tetroxide towards aromatic rings is highly dependent on the nature of its substituents. In general, substituted aromatic rings are readily oxidized to carbon dioxide when an electron donating groups such as alkoxy or hydroxy are present. Aromatic rings with electron withdrawing substitutents, such as $-CO_2H$, $-CO_2R$, COR, or $-CH_2OR$, resist ring oxidation (Figure 6). The presence of nitro group slowed or stopped oxidation, as well as the presence of *p*-carbometoxy substituent [19]. Aromatic rings of intermediate electron density, such as those attached to both an electron withdrawing and an electron donating group, display intermediate reactivity towards ring attack. Benzyl alcohols are oxidized rapidly by ruthenium tetroxide to the corresponding benzoic acid derivatives [21]. The reaction proceeds in high yield without degradation of benzene ring. Benzyl aryl ethers are oxidized, in low yields, to aryl benzoates, Some aromatic steroids also undergo benzylic oxidations [22,23].

Naphthalenes and other polycyclic aromatic compounds are oxidized in similar manner. Substitutents exert a substantial effect on the oxidation. A ring that is activated by an electron donating substituent is oxidized and a ring that is deactivated by an electron withdrawing substituent is preserved [24-26].

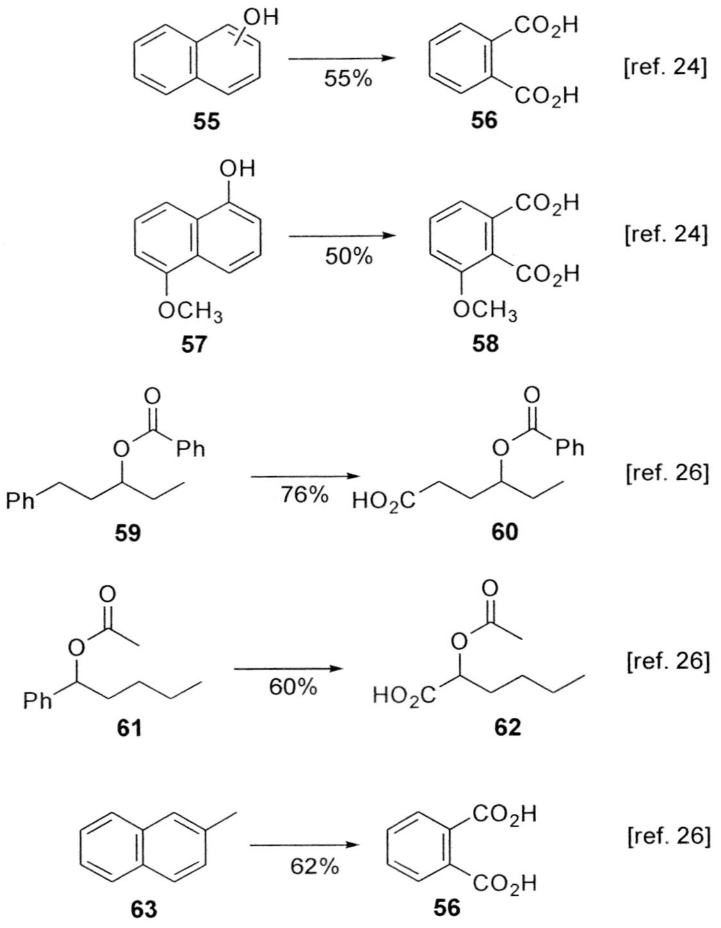

Figure 6.

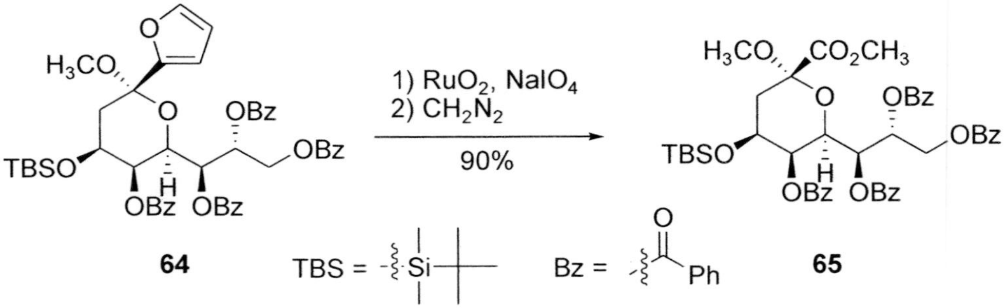

Furan rings are rapidly oxidized to the corresponding carboxylic acids [27].

2.5. Alcohols

Alcohols are highly reactive towards ruthenium tetroxide. Oxidation is often completed within seconds and the yields are usually high. Primary alcohols are oxidized to the corresponding carboxylic acids. Formation of esters, a common side reaction in some other oxidations of primary alcohols (e.g. chromate oxidations), is not observed [29]. Secondary alcohols are oxidized to the corresponding ketones. Vicinal diols are cleaved to the corresponding carbonyl compounds. Ruthenium tetroxide oxidizes allylic alcohols with loss of one or more carbon atoms. This transformation can be done with OsO4 and NaIO4 in lower yields [30]. Allylic alcohols are highly reactive and rather complex molecules can be oxidized to give a high yield of the corresponding carbonyl compounds (Figure 7) [31].

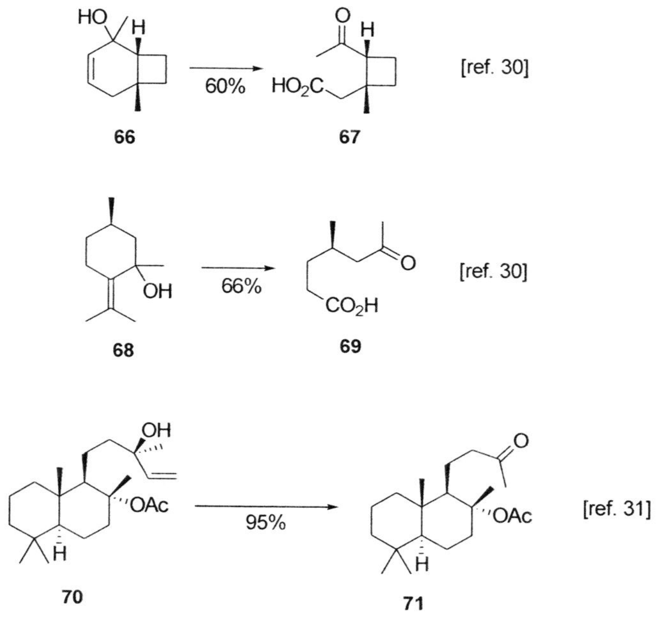

Figure 7.

Ruthenium tetroxide oxidation of alcohols does not affect □-lactone functionalities, which underwent ring opening and oxidation when treated with other oxidizing agents (Figure 8) [32,33].

As ruthenium tetroxide oxidizes both aromatic compounds and alcohols, it is important to note that an alcohol is oxidized preferentially to a phenyl ring (Figure 9) [8]. Alcohols are also oxidized preferentially to ethers, including more reactive benzyl ethers [34]. If an alcohol functionality is in the benzylic position the resulting, electron-withdrawing, carbonyl group protects the phenyl ring from oxidation [35,36].

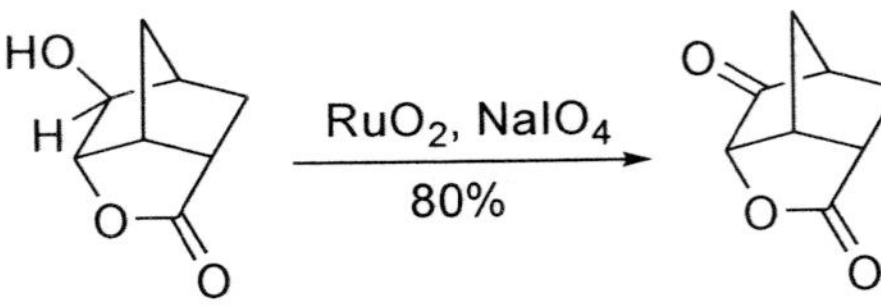

72 73 [ref. 32]

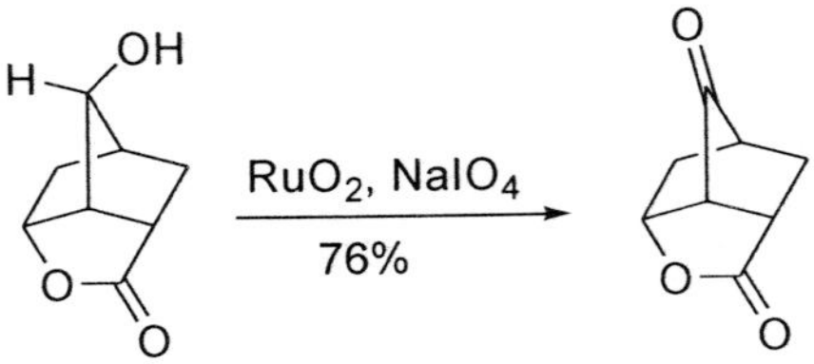

74 75 [ref. 33]

Figure 8.

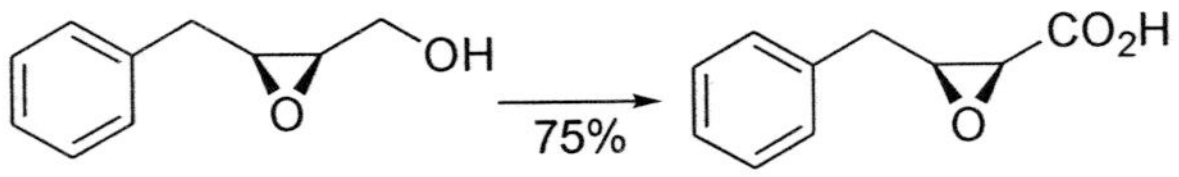

76 77 [ref. 8]

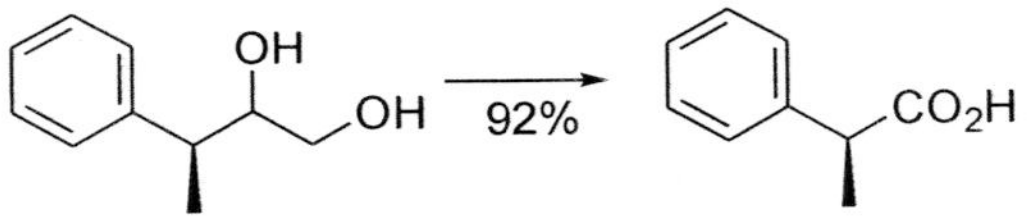

78 79 [ref. 8]

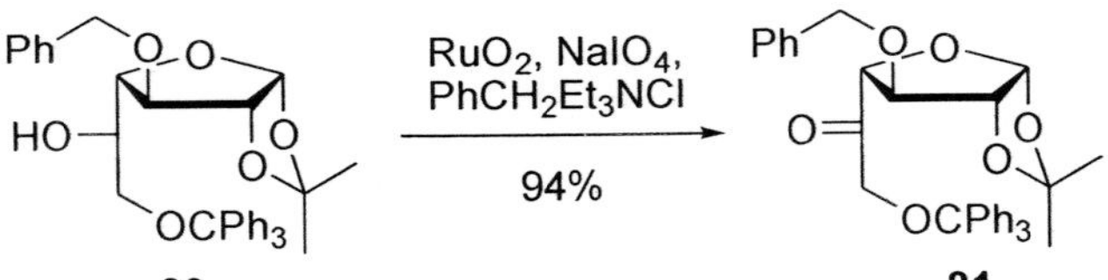

80 81 [ref. 34]

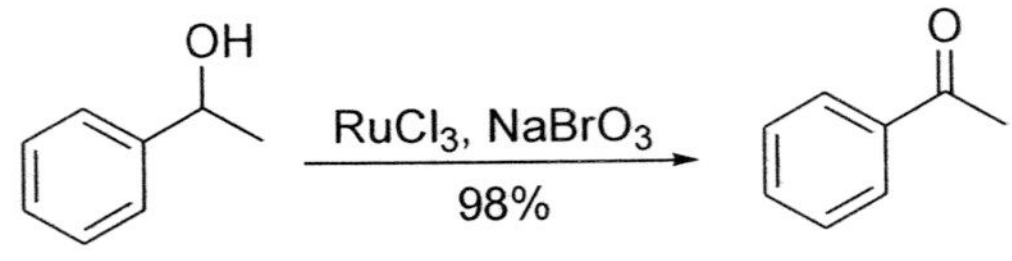

82 83 [ref. 35]

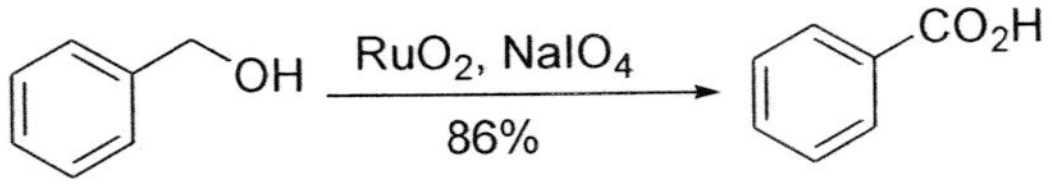

84 35 [ref. 36]

Figure 9

2.6. Aldehydes and Ketones

Aldehydes are oxidized to the corresponding carboxylic acids [36]. Ketones, in general, are resistant to ruthenium tetroxide oxidation. However there are reports that, in certain cases, ketones are oxidized with insertion of oxygen to give the corresponding lactones in a manner similar to a Beyer-Villiger oxidation [37]. α,β-Unsaturated ketones are oxidized to the corresponding carbonyl compounds with a loss of a carbon atom (Figure 10) [30].

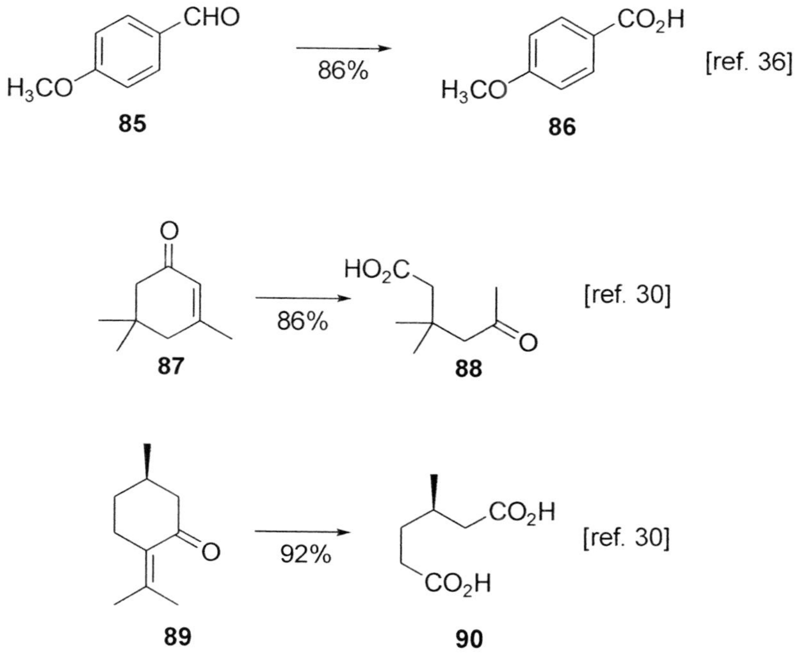

Figure 10.

2.7. Ethers

While a number of reagents are available that cleave ethers, very few agents are capable of oxidizing an ether to give a single product in a good yield. Most oxidizing agents do not oxidize ethers, oxidize only activated ethers, provide low yields of the oxidation products, or give mixtures of products. For example, permanganate oxidizes benzyl ethers to the corresponding carboxylic acids in the presence of a phase transfer catalyst [38]. Ruthenium tetroxide is one of the few oxidizing agents that oxidizes ethers. It oxidizes aliphatic ethers to esters, and cyclic ethers to lactones, in good to excellent yields (Figure 11). Yields of lactones are sometimes modest due to hydrolysis to the hydroxyacid form, which undergoes further oxidation to the corresponding ketoacid or a dicarboxylic acid [39,40]. If acetonitrile modification is used, hydrolysis of aliphatic esters is barely detectable and they are the major products [8]. However, small cyclic ethers, such as tetrahydrofuran, still give only modest yields of the corresponding lactones unless the reaction conditions are carefully controlled and optimized for preparation of the lactone [36,41]. Esters are resistant to further oxidation and attempts to carry the oxidation of esters to acid anhydrides were unsuccessful. Benzyl ethers are considerably more reactive compared to aliphatic ethers and are oxidized exclusively at the benzylic position to give benzoate esters [42]. Highly strained oxetanes are oxidized into the corresponding β-propiolactones in a low yield [43].

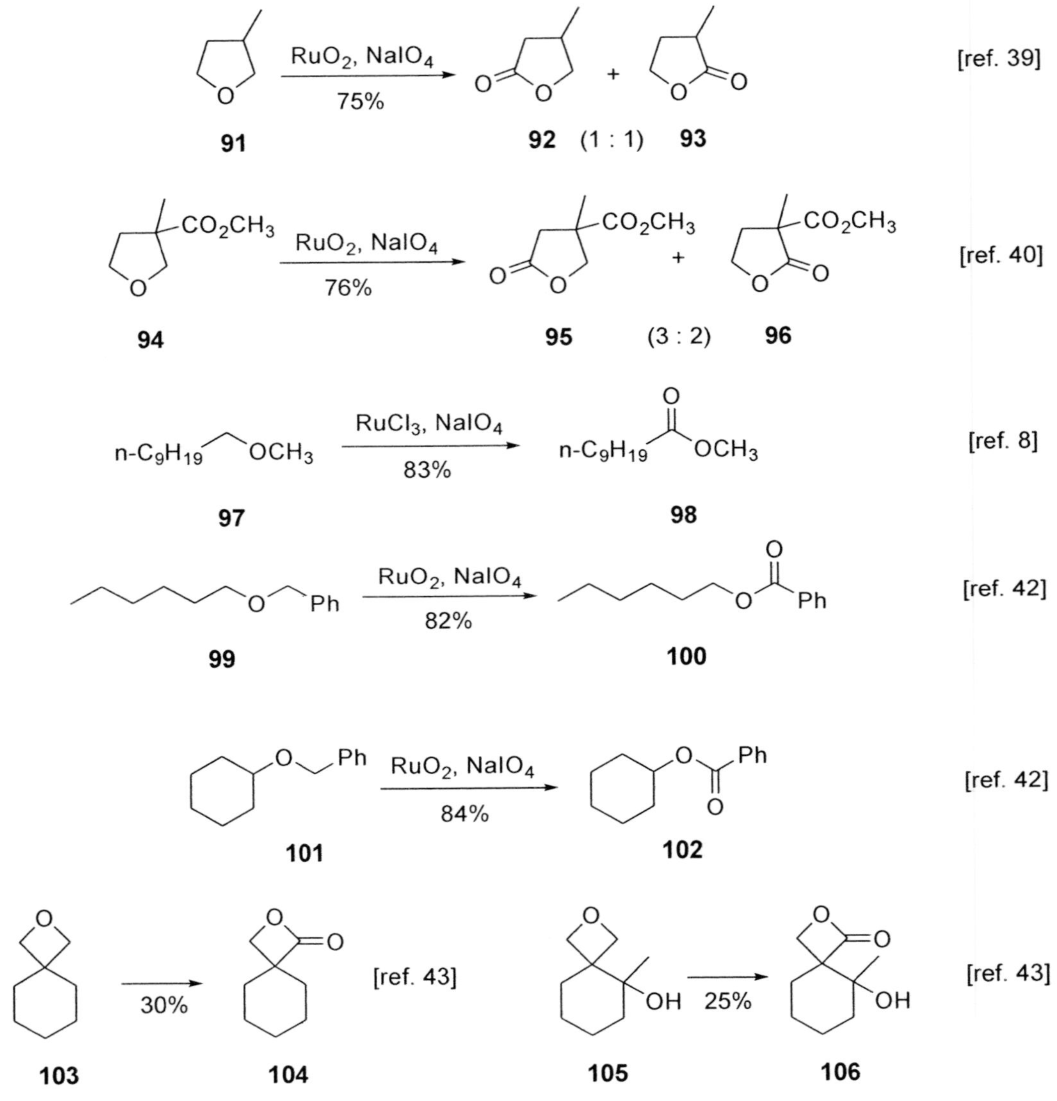

Figure 11.

2.8. Amines

Amines are rapidly oxidized by ruthenium tetroxide. However, mono- and disubstituted amines produced only complex mixtures of products. Tertiary amines are oxidized to amides or lactams [44,45]. The results are considerably better if there is an electronegative substituent on the nitrogen atom [46,47]. Thus, unlike esters, which cannot be oxidized to anhydrides, amides are easily oxidized to imides (Figure 12).

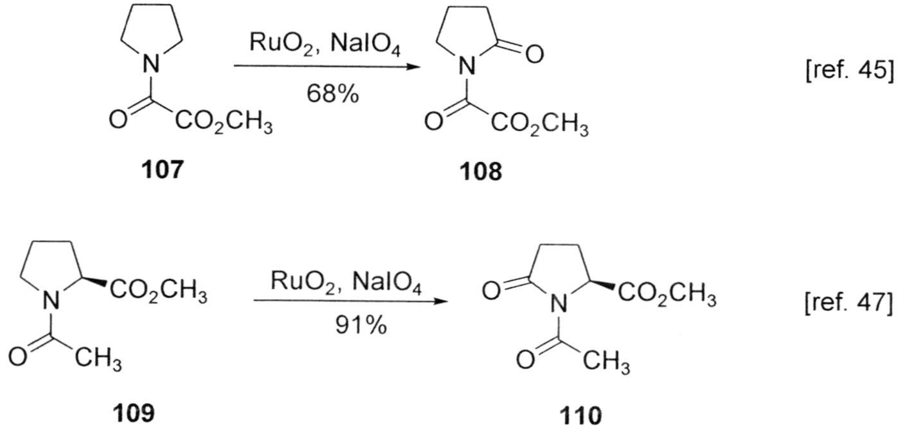

Figure 12.

2.9. Sulfur-containing Organic Compounds

Unlike ethers, amines or amides, where the methylene adjacent to the heteroatom is oxidized, organic sulfides are oxidized by ruthenium tetroxide at the heteroatom itself. The sulphoxide is usually a reaction intermediate, and it is oxidized further to the corresponding sulphone (Figure 13) [5,48-50].

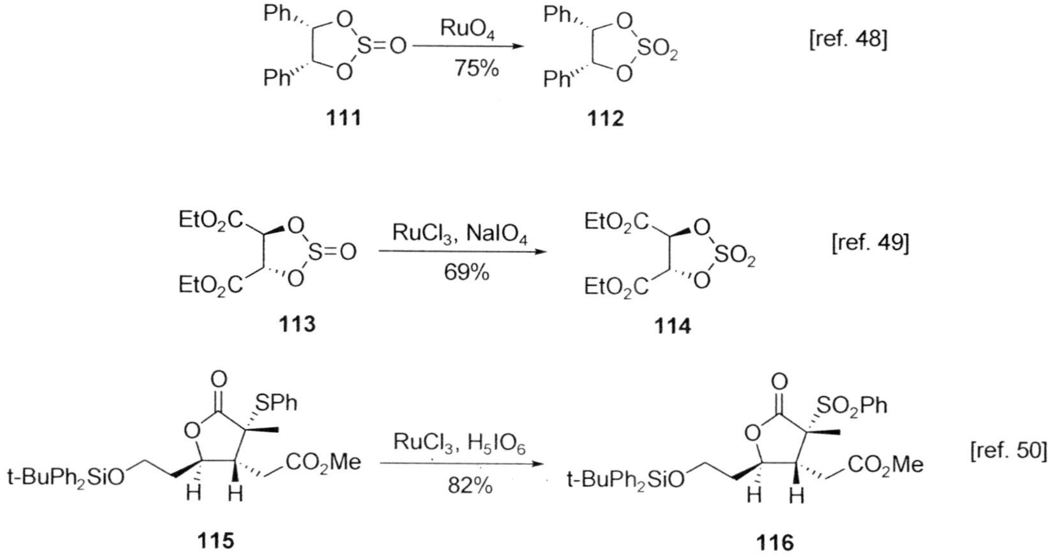

Figure 13.

3. Degradation Reactions

Degradation reactions were originally introduced into organic chemistry as a means of converting a natural product into a more valuable compound (e.g. Barbier-Wieland degradation of steroid side chain). Development of modern synthetic methods made such degradation methods obsolete. Another use for degradation methods was as a means of structural elucidation of natural product. Today, analytical techniques, such as NMR, are more suitable for structure elucidation of small molecules. However, new application of chemical degradation has gained more importance. Degradation methods are now being used for investigation of insoluble complex macromolecules. Over the past two decades, catalytic ruthenium tetroxide oxidation has been extensively used in structural studies of kerogen.

Prior to application in structural studies of kerogen, ruthenium tetroxide oxidations have been used in structural studies of other of organic macromolecules including coal [51-54], bitumen [55-58] and crude oil [59].

More recently it has been used in structural studies of algaenans [60,61], macromolecular fraction of soil [62], an amber polymer [63] and organic material in a meteorite [64].

4. Oxidative Ruthenium Tetroxide Degradation of Kerogen

Earliest studies on ruthenium tetroxide oxidation of kerogen were reported by Boucher et al. [65-67]. Their initial study was on Kimmeridge kerogen [65] and was expanded to Green River, Maoming, Messel kerogens and Loy Yang coal [66]. A catalytic ruthenium tetroxide degradation was applied both on kerogen concentrate and directly on raw kerogen.

Products of oxidation of Kimmeridge kerogen were C_4-C_{24} α,ω-dicarboxylic acids (51%) and C_9-C_{27} n-monocarboxylic acids (14%), C_{14}-C_{21} acyclic isoprenoid acids (16%), C_5-C_{12} branched dicarboxylic acids (7%), and cyclic acids (10%), which is an indication that this kerogen is composed mainly of linear aliphatic chains with a small contribution from cyclic structures [65]. It is interesting that n-monocarboxylic acids obtained in the course of oxidation of Kimmeridge kerogen exhibited maxima at C_{15} and C_{17}, while most other kerogens exhibit maxima at C_{16} and C_{18}. A relatively high co-oxidant/substrate ratio (20:1) was applied, which apparently resulted in complete degradation of aromatic structures. The results were complemented with solid state ^{13}C NMR analysis, which provided information about aromatic structures.

An extensive study of ruthenium tetroxide oxidation of Messel kerogen utilizing various co-oxidant/substrate ratios (from 4:1 to 50:1), reaction temperatures and times was done in order to maximize the yield of isolated degradation products [67]. Major degradation products (80% of the total isolated oxidation products) were C_4-C_{30} α,ω-dicarboxylic acids and C_7-C_{32} n-monocarboxylic acids along with some isoprenoid acids (C_{13} – C_{17} and C_{19} – C_{21}), branched mono- (C_{13} – C_{17}) and di- (C_5 – C_{11}) carboxylic acids and hopanoic acids (C_{30} – C_{32}). Distribution of the products depended on the ration of co-oxidant/substrate. Only trace quantities of aromatic acids were isolated. Similarly to Kimmeridge kerogen, the results indicate presence of mainly of linear aliphatic chains. Composition of oxidation products suggests that this kerogen is of algal origin with some contribution from terrestrial material.

Boucher et al. next reported a comparative study of several different kerogens and Loy Yang coal, which were representative of kerogens of various types (I, I/II, II and III) [66]. Straight chain carboxylic acids were the predominant products, ranging from 65% to 87%. All of the kerogens yielded n-monocarboxylic acids of a similar range with a maximum at C_{16}, with an exception of Kimmeridge kerogen whose n-monocarboxylic acids exhibited a maximum at C_{17}. Kerogens with a terrestrial contribution also showed presence of higher molecular weight acids ($C_{24} - C_{32}$). α,ω-Dicarboxylic acids showed considerable variety among the five samples and their precursors may be unique to each individual kerogen.

Standen and co-workers further improved their method by adapting it to a microscale and optimizing it with respect to co-oxidant/substrate ratio, reaction temperature, time, method of esterification of degradation products and use of suitable internal standard [68]. The method was tested on Messel and Kimmeridge clay kerogens as well as model compounds. The oxidant was changed from $RuCl_3$-$NaIO_4$ to a more reproducible RuO_2-H_5IO_6 system. The improved method was one of the analytical techniques used in structural elucidation of Alum Shale kerogen [69]. Ruthenium tetroxide degradation of Alum Shale kerogen yielded n-monocarboxylic acids, branched monocarboxylic acids, α,ω-dicarboxylic acids, branched dicarboxylic acids, which combined accounted for more than 90% of the isolated products. Monoaromatic di- and tri-carboxylic acids and aliphatic hydrocarbons were minor products of the degradation.

Ruthenium tetroxide oxidation was applied to two co-existing lithotypes of German brown coal. Lignin-derived aromatic acids were observed in the "woody" kerogen material and higher plant liptinite derived acids were more prominent in the "amorphous" kerogen sample, which is an indication of different origins for the two fossil fuel types [70].

Following pioneering work by Standen and coworkers, improvements by other authors were introduction of various degradations sequences involving ruthenium tetroxide as well as use of ruthenium tetroxide in combination with other analytical methods. Gelin and co-workers combined hydroiodic acid and ruthenium tetroxide degradations in their investigation of algaenans in recent and ancient marine kerogens [71]. Earlier studies have indicated presence of ether bonds in kerogens. Hydroiodic acid treatment cleaved ethereal bonds, to the corresponding alkyl iodides, while ruthenium tetroxide oxidized ethers to the corresponding esters. Comparison of the two different methods allowed ether functionalities to be characterized with a considerable degree of certainty. Presence of $C_{24} - C_{34}$ ω-17- and ω-18-oxoacids as well as $C_{24} - C_{34}$ ω-17,18-dioxoacids was interpreted as originating from the corresponding ether linkages in the original kerogen. The main oxidation products were $C_8 - C_{18}$ n-alkanedioic acids and $C_{12} - C_{18}$ n-alkanoic acids, which appear to originate from oxidation of ethers at mid-chain vicinal positions. Presence of ether bonds was confirmed by results of hydroiodic acid degradation of the same kerogen. The authors concluded that algaenans are selectively preserved and may serve as a source of n-alkanes in marine crude oils.

Chemical structures of several fresh water algae and three different kerogens were compared by means of ruthenium tetroxide degradation [72]. By applying ruthenium tetroxide oxidation is was possible to compare details of chemical structure of the extant and fossil algaenans. RuO_4 oxidation products provided a chemical signature that allowed identification of algal family that was the main contributor to the organic matter of each kerogen. Comparison of RuO_4 degradation products of the Messel and Fuquene kerogens and

Tetraedron minimum and *Pediastrum boryanum*, respectively, showed that the building blocks of each pair were virtually identical. Thus, the fossil algaenans were preserved without major chemical changes. A difference was observed in the distribution of the long-chain dicarboxylic acids, which may be due to the differences in growth conditions between *T. minimum* that contributed to Messel kerogen and the one that was used in the study of the extant algaenan. Another difference was the broader distribution of dicarboxylic acids in the product mixtures of kerogens compared to extant algaenans, which probably reflects a difference in the amount of double bonds and ether-linkages of those biopolymers. In addition, fossil algaenans (Messel and Fuquene kerogens) showed a higher degree of ether cross-linking. RuO_4 oxidation of extant *Botryococcus Braunii* algaenan yielded C_7-C_{26} α,ω-dicarboxylic, C_{17}-C_{33} oxo- and C_{27}-C_{34} dioxo-dicarboxylic acids. This is the first time that dioxo-dicarboxylic acids have been observed among the RuO_4 oxidation products of algaenans. Products of RuO_4 oxidation Fuquene kerogen contained C_7-C_{24} α,ω-dicarboxylic and shorter chain oxo-dicarboxylic acids in a lower concentration. No dioxo-dicarboxylic acids were observed. Some of the observed differences were explained by the presence of *Coelastrum reticulatum*, while others were attributed to removal of material on the periphery of the macromolecule in the course of diagenesis.

Ruthenium tetroxide oxidation of Estonian Kukersite and Guttenberg Member kerogen microfossil produced mono-, di- and tri-carboxylic aliphatic acids [73]. A considerable number of oxocarboxylic acids of rather diverse structures were identified among the degradation products. Major degradation products of Estonian Kukersite were dicarboxylic acids while monoacids were predominant products of degradation of Guttenberg Member kerogen. It is interesting that, among the degradation products of the same kerogen, a series (C_6 – C_{19}) of methyl ketones was identified. It appears that precursor of this kerogen are selectively preserved algaenan cell walls of an extinct algae *Gloecapsomorpha prisca*.

Solid state ^{13}C NMR and FTIR study of two type I kerogens from the Dead Sea indicated a large contribution of *B. Braunii* algaenan [74,75]. Extractable organic matter from those kerogens was composed predominantly of organo-sulfur compounds derived from *B. Braunii*. Chemical degradation sequence was improved to include an alkaline hydrolysis prior to oxidation and ruthenium tetroxide oxidation was done under ultrasound conditions. Alkaline hydrolysis released a C_{16} – C_{34} series of fatty acids with even/odd predominance. Ruthenium tetroxide oxidation gave C_{16} – C_{34} fatty acids as the major products also with an even/odd predominance and distribution similar to that of hydrolysis products. However, oxidation did not reveal presence of compounds characteristic of *B. Braunii* algaenans such as α,ω-dicarboxylic acids and oxoacids, which lead authors to conclude that composition of those kerogens is not associated with *B. Braunii*. It is possible that *B. Braunii* that was involved in formation of those kerogens exhibited a high lipid content, which gave rise to biomarkers isolated in the course of solvent extraction, and low algaenan contend, which resulted in its negligible contribution to the sedimentary organic matter.

Ruthenium tetroxide oxidation has been developed into a method for chemical fingerprinting of both extant algaenans as well as algaenans from ancient sediments [76]. The authors have shown that composition of the degradation products is species specific and that GC fingerprint of RuO_4 degradation products is preserved in sedimentary organic matter. A review of algae and plant biomolecules and their fossil analogues has recently appeared [77].

Figure 14 shows a general structure of an algaenan as established by de Leeuw and co-workers.

Ruthenium tetroxide degradation, along with pyrolysis-GC, of Albian black shale reveled yet another source of ancient sedimentary organic matter [78]. The presence of ether lipids among the degradation products was characteristic of marine planktonic archaea. Apparently, in the course of diagenesis, archaea lipids polymerized into macromolecules producing most of the Albian black shale organic matter.

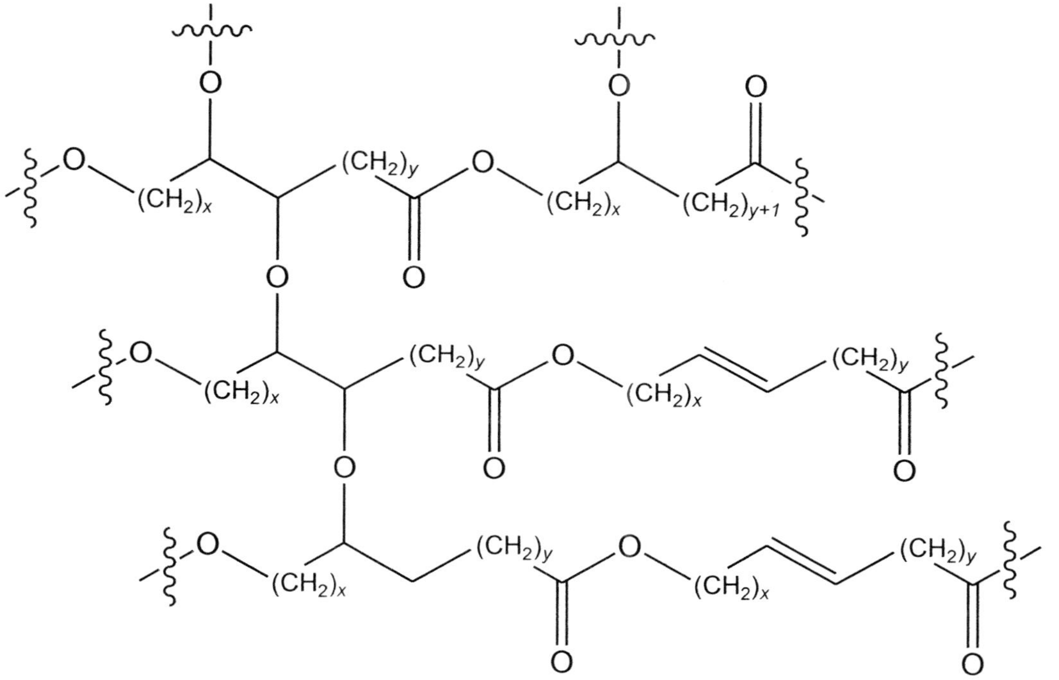

Figure 14.

Reiss and co-workers isolated several series of new hopanoid triterpenes after ruthenium tetroxide oxidation and esterification of the oxidation products of Messel shale kerogen (Figure 15) [79]. The authors proposed that hopanoids originated from incorporation of bacterial material and are bound to aromatic subunits of kerogen. The major oxidation products were n-monocarboxylic and α,ω-dicarboxylic acids. A trace amounts of aromatic acids were also identified.

Kirbii and co-workers combined hydrolysis under phase transfer conditions with ruthenium tetroxide oxidation to analyze Maroccan Timahdit oil shale kerogen [80]. By applying a multi-step alkaline hydrolysis of the kerogen concentrate in the presence of 18-crown-6 as a phase transfer catalyst the authors have been able to solubilze 65% of the initial kerogen. High degree of solubilization is an indication of presence of considerable amount of ester groups in this kerogen. Composition of hydrolysis products from the various stages of hydrolysis indicated a kerogen with a mainly aromatic core and predominantly aliphatic periphery. The main hydrolysis products were aliphatic and aromatic monocarboxylic acids, various dicarboxylic acids including a relatively large amount of branched dicarboxylic acids, and aliphatic and aromatic alcohols. The same kerogen was subjected to a catalytic ruthenium tetroxide oxidation. The amount of solubilized kerogen was not reported. Distribution of aliphatic acids resembled that of hydrolysis products indicating that some esters have been hydrolyzed in the course of ruthenium tetroxide treatment. While the composition of monocarboxylic acids isolated by hydrolysis and ruthenium tetroxide oxidation was similar, even/odd carbon number predominance among the monocarboxylic acids was considerably less pronounced compared to monocarboxylic acids isolated in the course of hydrolysis. This may reflect contribution of alkyl substituents on the aromatic rings. Composition of α,ω-dicarboxylic acids isolated in the course of ruthenium tetroxide oxidation was very different from those isolated in the course of hydrolysis, with C_7 and C_9 carbon number being the most predominant (C_{11} max for hydrolysis). Also, a larger amount of longer chain (C_{17}-C_{26}) was isolated. Main new class of products, which were not observed in the course of hydrolysis, were hopanoic acids. As hopanoic acids were isolated only in the course of ruthenium tetroxide oxidation, it was assumed that they were bound to the kerogen matrix by ether groups. Ether functionalities of this kerogen have also been studied by methods other than ruthenium tetroxide oxidation [81].

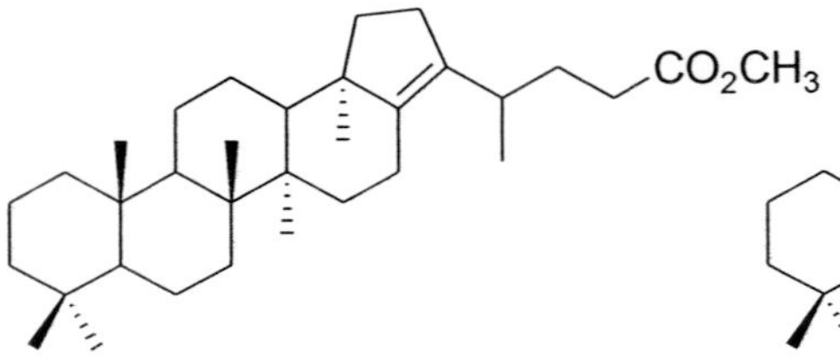

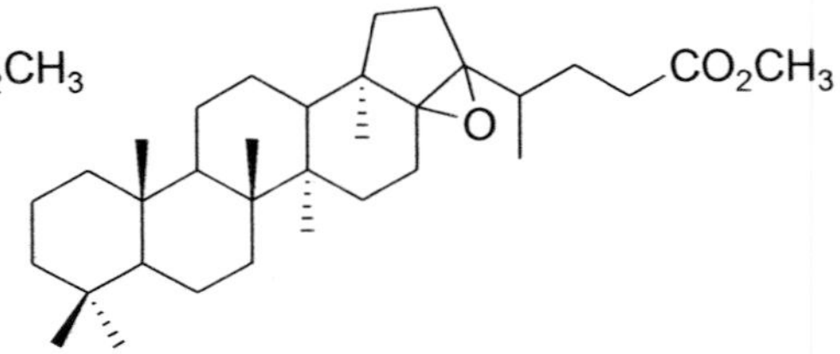

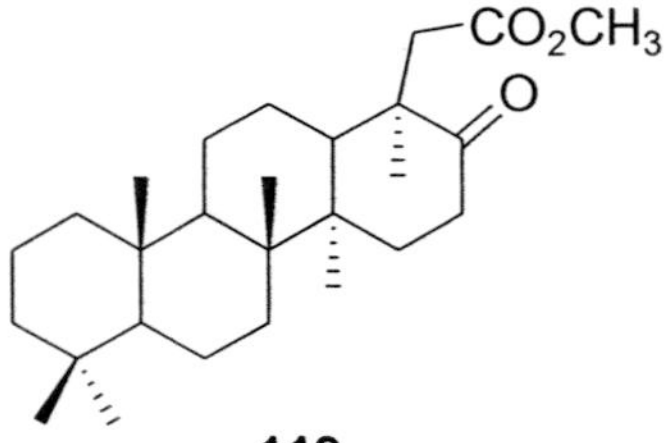

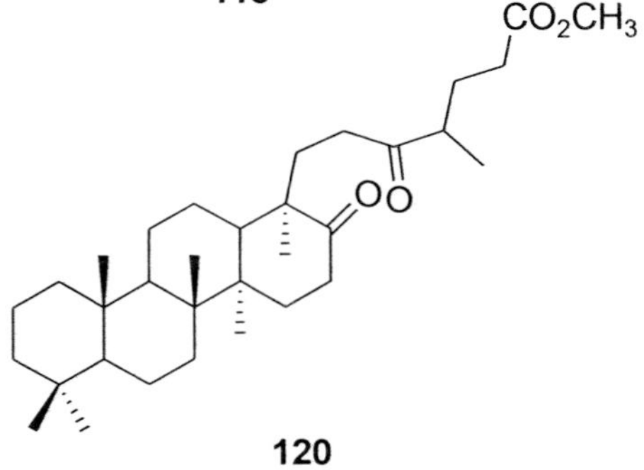

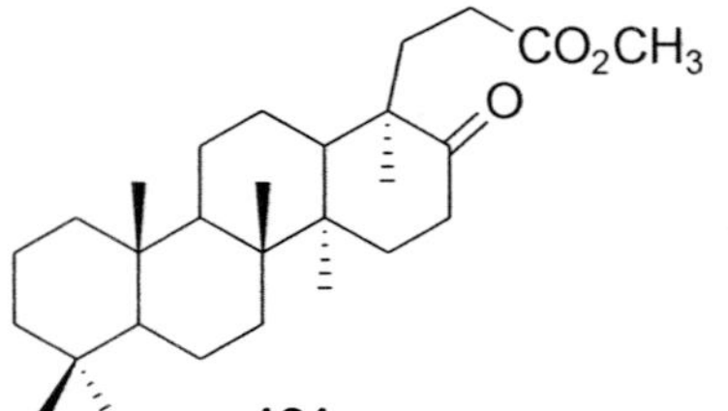

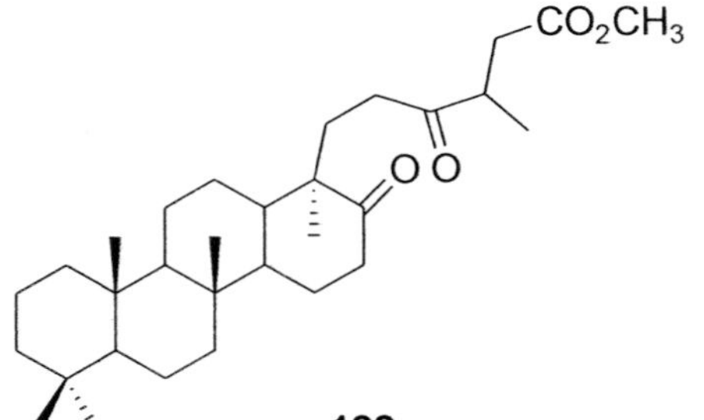

Figure 15.

A degradation sequence, comprising of reduction of kerogen with Ni(0)cene-LiAlD$_4$ reagent, BCl$_3$ – mediated ether cleavage and finally ruthenium tetroxide oxidation of the residue was introduced by Richnow and co-workers (Figure 16) [82]. In the first step Ni(0)cene [bis(1,5-cyclopentadiene)nickel(0)]-LiAlD$_4$ reagent cleaved sulfur bonds leaving a deuterium labeled carbon atom. Next, ether and ester bonds were cleaved by boron trichloride. As ethers are cleaved to an alcohol and an alkyl chloride, the latter was converted into the corresponding alkane by reduction with LiAlH$_4$. Finally, the residue, believed to be composed mainly of aromatic structures, was oxidized with ruthenium tetroxide. A comparative study of sulfur-rich Monterey kerogen and Monterey oil by this method revealed, as expected, a high degree sulfur linkages in both. Major products of the degradation sequence were linear aliphatic compounds, which are attached by sulfur or oxygen, or both simultaneously, to the macromolecular network by or are bonded to aromatic structures of the macromolecule. A number of biomarkers were identified among the degradation products including acyclic isoprenoids, triterpenoids, hopanoids and steroids. The steroid A-ring was identified as a principal site of its bonding to an aromatic unit. Presence of C$_{37}$ and C$_{38}$ *n*-alkanes among the desulfurization products indicated contribution of lipids from algae, while presence of hopanoids indicated that a portion of the bacterial membrane constituents is incorporated into the macromolecular networks without any significant alteration.

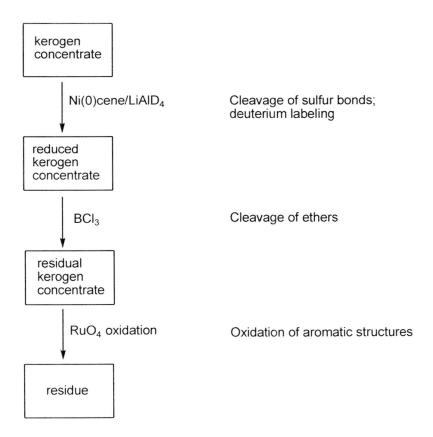

Figure 16: Ni(0)cene/LiAlD$_4$ reduction – BCl$_3$ ether cleavage – RuO$_4$ oxidation sequence (after Figure 2., ref. 81).

A catalytic ruthenium tetroxide oxidation in conjunction with hydrolysis was employed in order to study ether and ester functionalities of Aleksinac kerogen by means of hydrolysis – ruthenium tetroxide oxidation – hydrolysis degradation sequence (Figure 17) [83]. Since RuO_4 oxidizes ethers to esters kerogen concentrate was hydrolyzed first with methanolic potassium hydroxide and soluble products were isolated. Hydrolyzed kerogen concentrate was subjected to a ruthenium tetroxide catalyzed oxidation according to a modified Sharpless procedure [8]. Soluble products were isolated, converted to methyl esters by treatment with diazomethane. The remaining residue, which contained new esters, formed by ruthenium tetroxide oxidation of ethers, was hydrolyzed with methanolic potassium hydroxide and the soluble products were isolated and analyzed. The amount of co-oxidant was adjusted so that the combined yield of the three degradation steps was maximized. As considerable amount of kerogen concentrate was solubilized in the two hydrolyses, co-oxidant/substrate ratio was considerably smaller compared to other degradation methods that involve ruthenium tetroxide (~7 vs. ≥20). Major products of the initial hydrolysis of kerogen concentrate were various aromatic acids, $C_{10} - C_{24}$ n-monocarboxylic acids and $C_6 - C_{18}$ α,ω-dicarboxylic acids along with a some α-methylmonocarboxylic acids. Yield of hydrolysis products was 10.3% based on the initial kerogen. More than 50% of the remaining, ester-free kerogen, was oxidized by ruthenium tetroxide. Soluble products (35.4% yield) were separated and kerogen concentrate subjected to a second hydrolysis. Major products were $C_{12} - C_{28}$ n-monocarboxylic acids (31.7%) and $C_7 - C_{25}$ α,ω-dicarboxylic acids (29.5%), aromatic acids (14.3) along with some branched acids (~2.2%). Some dihydroabietic acid was isolated as well as a trace amount of oleic acid. Although the products of the second hydrolysis are supposed to reflect the original ether structures in Aleksinac kerogen, it is likely that some of them were esters originally trapped in the kerogen core, while others may have been ester groups inside kerogen's matrix that have been inaccessible to hydroxide in the original hydrolysis and become accessible after its degradation.

Ruthenium tetroxide oxidation of hydrolyzed kerogen concentrate gave a high yield (68%) of soluble products. Among the aliphatic products, two new series of oxodicarboxylic acids and one series of oxomonocarboxylic acids were isolated [84], which may indicate a contribution of algal precursors to this kerogen.

The same, hydrolysis – ruthenium tetroxide oxidation – hydrolysis, sequence was applied to Messel kerogen [85]. It is interesting to compare the results with other studies of Messel kerogen by ruthenium tetroxide oxidation [67,72]. Main degradation products were n-monocarboxylic and α,ω-dicarboxylic acids. However, as less RuO_4 was used (co-oxidant/substrate ratio ~6) and considerable portion of aromatic structures was preserved. Model studies indicate that, although initial oxidation of an aromatic ring is somewhat slower compared to functionalities such as alcohols, alkenes and alkynes, once the oxidation of a ring starts its degradation is very rapid [86]. Furthermore, degradation of a single benzene ring uses up considerable amount of the co-oxidant. Thus, when a limited amount of the co-oxidant is used, a large portion of aromatic structures may be unaffected by it. Among the degradation products of both Aleksinac and Messel kerogen alkylaromatic acids (benzenic and naphthenic) were observed. The total yield of aromatic acids (35%) obtained in this degradation sequence most likely overestimates actual proportion of the aromatic structures. Some "aromatic" acids are actually alkylaromatic and, thus, some aliphatic portions are included in the 35% yield. Furthermore, as the yield of the soluble products was 64.6% they represent only a part of this kerogen's structure.

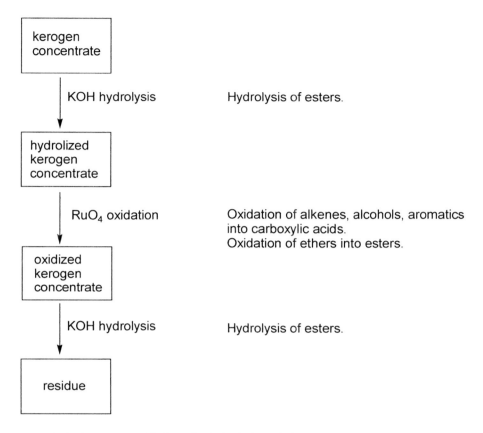

Figure 17: Hydrolysis – RuO$_4$ oxidation – hydrolysis degradation sequence.

Interestingly, oxodicarboxylic acids, which were prominent among the products of ruthenium tetroxide oxidation of this kerogen by other groups [72] were not observed. This can be rationalized by a relatively low reactivity of ethers, which is lower than that of aromatic compounds [87]. Only a limited amount of the co-oxidant was employed and apparently it was used up on oxidation of other functionalities.

Yoshioka and Ishiwatari optimized Standen's ruthenium tetroxide oxidation of kerogen [67] to recover low-molecular weight acids as well as aromatic acids and applied it to several marine and lucustrine kerogens as well as model compounds [86]. They isolated $C_2 - C_{26}$ α,ω-dicarboxylic acids, $C_4 - C_7$ branched dicarboxylic acids, $C_2 - C_{28}$ n-monocarboxylic acids and benzenoic acids as the main degradation products of both marine and lacustrine kerogens. There was a significant difference in composition of α,ω-dicarboxylic acids between the marine and lacustrine kerogens. Marine kerogens showed a maximum at C_2 and a decreasing trend with increasing carbon number. It was proposed that precursors of the short chain α,ω-dicarboxylic acids are melanoidins and unsaturated fatty acids. Lacustrine kerogens exhibited a maximum at C_9 whose most likely precursors are Δ^9 unsaturated fatty acids. One of the lacustrine kerogens also exhibited maxima at C_{21} and C_{23} probably due to incorporation of aliphatic biopolymers present in algal cell wall. It is important to note that C_3

α,ω-dicarboxylic acid was not isolated presumably because it is easily oxidized by ruthenium tetroxide. Authors also suggested that carbon preference index (CPI) of *n*-carboxylic acids isolated in the course of ruthenium tetroxide degradation may be used to estimate the extend of diagenesis of a kerogen.

Study of Jinghan asphaltenes and kerogens by ruthenium tetroxide oxidation combined with other degradation methods (nickel boride desulfurization and boron tribromide ether and ester cleavage) showed considerable structural similarities between the asphalten and kerogen [88]. The major products of ruthenium tetroxide oxidation were *n*-carboxylic acids, α,ω-dicarboxylic acids, isoprenoid acids, steranoic acids and hopanoic acids and were assumed to originate from substituents on the aromatic structures.

Li and coworkers reported study of kerogen from Hongshuizhuang Formation by ruthenium tetroxide oxidation [89]. The main degradation products were C_5-C_{19} *n*-monocarboxylic acids, C_5 – C_{30} branched monocarboxylic acids, C_6 – C_{16} α,ω-dicarboxylic acids, C_3 and C_4 branched dicarboxylic acids, benzene mono- to hexa-carboxylic acids, naphthenic acids, cyclohexanecarboxylic acid and 4-methylcyclohexanecarboxylic acid. The results indicate that short polymethylenic chains and terminal alkyl group are the main structural features of this kerogen. Based on structural similarities of this kerogen and cyanobacterial resistant material, the authors concluded that cyanobacteria were the most likely precursor of this kerogen.

5. CONCLUSION

Ruthenium tetroxide's selectivity and its ability to oxidize organic compounds under mild conditions make it a valuable tool for structural elucidation of kerogen particularly when combined with other means of structural elucidation of kerogen. A number of research groups made effective use of either a combination of, or a sequence of, different degradation methods in structural studies of sedimentary organic matter. Ruthenium tetroxide allowed study of ether functionalities in kerogen and their differentiation from esters, two oxygen-containing functionalities that are difficult to both investigate and distinguish from each other by other methods. An important development has been an application of ruthenium tetroxide oxidation to chemically fingerprint extant and fossil algaenans. Further optimization of degradation procedures is needed as well as further investigation of reactivity of ruthenium tetroxide under a more realistic degradation conditions. Model compounds are usually studied under optimal conditions and it is assumed that the reactivity of structural elements of kerogen would be identical to that of individual molecules in a solution. There is a continuous need to study ruthenium tetroxide oxidation beyond the simplistic reliance on knowledge of general reactivity of ruthenium tetroxide towards organic functional groups. Such studies are likely to benefit not only research on sedimentary organic matter, but also organic geochemistry in general as well other disciplines such as practical, synthetic and mechanistic organic chemistry.

REFERENCES

[1] Vandenbroucke, M. &argeau, C. (2007). *Geochem. 38*, 719-833.
[2] Durand, B. (Ed.), (1980). Kerogen, Insoluble Organic Matter from Sedimentary Rocks. Éditions Technip, Paris.
[3] Vitorović, D. (1980). Structure elucidation of kerogen by chemical methods. In: Durand, B. (Ed.), Kerogen, Insoluble Organic Matter from Sedimentary Rocks. Éditions Technip, Paris, pp. 301-338.
[4] Vitorovic, D.; Amblès, A. & Djordjevic, M. (1985). *Org. Geochem. 10*, 1119-1126.
[5] Djerassi, C. & Engle, R.R. (1953). *J. Am. Chem. Soc. 75*, 3838-3840.
[6] Courtney, J.L. & Swansborough, K.F. (1972). *Rev. Pure and Appl. Chem. 22*, 47-54.
[7] Courtney, J.L. (1986). In Organic Synthesis by Oxidation with Metal Compounds; Mijs, W.J.; de Jonge, C.R.H.I.; Eds.; Plenum Press; New York, NY, Chapter 8, pp. 445-467.
[8] Carlsen, P.H.J.; Katsuki, T.; Martin, V.S.; et al. (1981). *J. Org. Chem. 46*, 3936-3938.
[9] Carlsen, P.H.J. (1987). *Synth. Commun. 17(1)*, 19-23.
[10] Tenaglia, A.; Terranova, E. & Waegell, B. (1992). *J. Org. Chem. 57(20)*, 5523-5528.
[11] Bakke, J.M. & Lundquist, M. (1986). *Acta Chem. Scandinavica B 40*, 430-433.
[12] Hasegawa, T.; Niwa, H. & Yamada, K. (1985). *Chem. Letters*, 1385-1386.
[13] Coudret, J.L.; Zölllner, S.; Ravoo, B.J.; et al. (1996). *Tetrahedron Lett. 37(14)*, 2425-2428.
[14] Young, W.B.; Link, J.T.; Masters, J.J.; et al. (1995). *Tetrahedron Lett. 36(28)*, 4963-4966.
[15] Gopal, H. & Gordon, A.J. (1971). *Tetrahedron Lett.* 2941-2944.
[16] Zibuck, R. & Seebach, D. (1988). *Helvetica Chimica Acta, 71*, 237-240.
[17] Dragojlovic, V. unpublished results.
[18] Caputo, J.A. & Fuchs, R. (1967). *Tetrahedron Lett.* 4729-4731
[19] Kasai, M. & Zifer, H. (1983). *J. Org. Chem. 48*, 2346-2349.
[20] Danishefsky, S. & Harvey, D.F. (1985). *J. Am. Chem. Soc. 107*, 6647-6652.
[21] Ilsley, W.H.; Zingaro, R.A. & Zoller, Jr., J.H. (1986). *Fuel 65*, 1216-1220.
[22] Piatak, D.M. & Ekundayo, O. (1973). *Steriods* 475-481.
[23] Piatak, D.M.; Herbst, G.; Wicha, J.; et al. (1969). *J. Org. Chem. 34*, 116-120.
[24] Ayres, D.C. & Hossain, A.M.M. (1975). *J.C.S. Perkin I*, 707-710.
[25] Chakraborti, A.K. & Ghatak, U.R. (1983). *Synthesis* 746-748.
[26] Nuñez, M.T. & Martin, V.S. (1990). *J. Org. Chem. 55*, 1928-1932.
[27] Danishefsky, S.J.; De Ninno, M.P. & Chen, S.-h (1988). *J. Am. Chem. Soc. 110*, 3929-3940.
[28] Frye, S.V. & Eliel, E.L. (1985). *Tetrahedron Lett. 26(33)*, 3907-3910.
[29] Eaton, P.E. & Mueller, R.H. (1972). *J. Am. Chem. Soc. 94*, 1014-1016.
[30] Webster, F.X.; Rivas-Enterrios, J.; & Silverstein, R.M. (1987). *J. Org. Chem. 52*, 689-691.
[31] Martres, P.; Perfetti, P.; Zahra, J.-P.; et al. (1994). *Tetrahedron Lett. 35(1)*, 97-98.
[32] Moriarty, R.M.; Gopal, H. & Adams, T. (1970). *Tetrahedron Lett.* 4003-4006.
[33] Gopal, H.; Adams, T. & Moriarty, R.M. (1972). *Tetrahedron 28*, 4259-4266.
[34] Morris Jr., P.E. & Kiely, D.E. (1987). *J. Org. Chem. 52*, 1149-1152.

[35] Yamamoto, Y.; Suzuki, H. & Moro-oka, Y. (1985). *Tetrahedron Lett. 26(17)*, 2107-2108.
[36] Cornely, J.; Su Ham, L.M.; Meade, D. E.; et al. (2003). *Green Chemistry 5*, 34-37.
[37] Nutt, R.F.; Arison, B.; Holly, F.W.; et al. (1965). *J. Am. Chem. Soc. 87*, 3273.
[38] Schmidt, H.J. & Schafer, H.J. (1979). *Angew. Chem. Int. Ed. 18*, 69-70.
[39] Smith III, A.B. & Scarborough Jr., R.M. (1980). *Synth. Commun. 10(3)*, 205-211.
[40] Scarborough Jr., R.M.; Toder, B.H. & Smith III, A.B. (1980). *J. Am. Chem. Soc. 102(11)*, 3904-3913.
[41] Gonsalvi, L.; Arends, I.W.CE. & Sheldon, R. A. (2002). *JCS Chem. Commun.* 202-203.
[42] Schuda, P.F.; Cichowicz, M.B. & Heimann, M.R. (1983). *Tetrahedron Lett. 24*, 3829-3830.
[43] Kato, M.; Kitahara, H. & Yoshikoshi, A. (1985). *Chem. Letters* 1785-1788.
[44] Bettoni, G.; Carbonara, G.; Franchini, C.; et al. (1981). *Tetrahedron 37(24)*, 4159-4164.
[45] Tangari, N.; Giovine, M.; Morlacchi, F.; et al. (1985). *Gazzeta Chimica Italiana 115*, 325-328.
[46] Sheehan, J.C. & Tulis, R.W. (1974). *J. Org. Chem. 39(15)*, 2264-2267.
[47] Yoshifuji, S.; Matsumoto, H.; Tanaka, K.-i.; et al. (1980). *Tetrahedron Lett. 21*, 2963-2964.
[48] Lowe, G. & Salamone, S.J. (1983). *J.C.S. Chem. Commun.* 1392-1394.
[49] Gao, Y. & Sharpless, K.B. (1988). *J. Am. Chem. Soc. 110(22)*, 7538-7539.
[50] Rodríguez, C.M.; Ode, J.M.; Palazón, J.M.; et al. (1992). *Tetrahedron 48(17)*, 3571-3576.
[51] Blanc, P. & Albrecht, P. (1991). *Org. Geochem. 17(6)*, 913-918.
[52] Stock, L.M. & Wang, S.H. (1989). *Energy and Fuels 3*, 535-536. and references cited therein.
[53] Mallya, N. & Zingaro, R.A. (1984). *Fuel 63*, 423-425.
[54] Satory, M.; Ken-ichi, U. & Masakatsu, N. (1994). *Energy and Fuels 8(6)*, 1379-1383.
[55] Mojelsky, T.W.; Montgomery, D.S. & Strausz, O.P. (1985). *AOSTRA J. Res. 2(2)* 131-137.
[56] Mojelsky, T.W.; Montgomery, D.S. & Strausz, O.P. (1986). *AOSTRA J. Res. 2(3)* 177-184.
[57] Mojelsky, T.W.; Montgomery, D.S. & Strausz, O.P. (1986). *AOSTRA J. Res. 3(1)* 43-51.
[58] Mojelsky, T.W.; Ignasiak, T.M.; Frakman, Z.; et al. (1992). *Energy and Fuels 6 (1)*, 83-96.
[59] Warton, B.; Alexander, B.; Kagi, R.I. (1990). *Org. Geochem. 30*, 1255-1272.
[60] Blokker, P.; Schouten, S.; van den Ende, H.; et al. (1998). *Org. Geochem.29(5-7)*, 1453-1468.
[61] Schouten, S.; Moerkerken, P.; Gelin, F.; et al. (1998). *Phytochemistry 49(4)*, 987-993.
[62] Quénéa, K.; Derenne, S.; Gonzalez-Villa, F.J.; et al. (2005). *Org. Geochem. 36*, 1151-1162.
[63] Anderson, K.B. (2001) The nature and fate of natural resins in the geosphere. Part XI. Ruthenium tetroxide oxidation of a mature Class Ib amber polymer. *Geochem. Trans.*, 3. http://www.geochemicaltransactions.com/content/pdf/1467-4866-2-21.pdf (accessed on September 23, 2007).

[64] Remusat, L.; Derenne, S. & François, R. (2005). *Geochimica et Cosmochimica Acta 69 (17)*, 4377-4386.
[65] Boucher, R.J.; Standen, G.; Patience, R.L.; et al. (1990). *Org. Geochem. 16*, 951-958.
[66] Boucher, R.J.; Standen, G. & Eglinton, G. (1991). *Fuel 70*, 695-702.
[67] Standen, G.; Boucher, R.J.; Rafalska-Bloch, J.; et al. (1991). *Chemical Geology 91*, 297-313.
[68] Standen, G. & Eglinton, G. (1992). *Chemical Geology 97*, 307-320.
[69] Bharati, S.; Patience, R.L.; Larter, S.R.; et al. (1995). *Org. Geochem. 23 (11/12)*, 1043-1058.
[70] Standen, G.; Boucher, R.J.; Eglinton, G.; et al. (1992). *Fuel 71(1)*, 31-36.
[71] Gelin, F.; Boogers, I.; Noordeloos, A.A.M.; et al. (1996). *Geochimica et Cosmochimica Acta 60(7)*, 1275-1280.
[72] Blokker, P.; Schouten, S.; de Leeuw, J.W.; et al. (2000). *Geochimica et Cosmochimica Acta 64(12)*, 2055-2065.
[73] Blokker, P.; Bergen, P.V.; Pancost, R.; et al. (2001). *Geochimica et Cosmochimica Acta 65*, 885-900.
[74] Grice, K.; Schouten, S.; Blokker, P.; et al. (2003). *Org. Geochem. 34*, 471-482.
[75] Grice, K.; Schouten, S.; Nissenbaum, A.; et al. (1998). *Org. Geochem. 28*, 195-216.
[76] Blokker, P.; van den Ende, H.; de Leeuw, J.W.; et al. (2006). *Org. Geochem. 37*, 871-881.
[77] de Leeuw, J.W.; Versteegh, G.J.M. & van Bergen, P.F. (2006). *Plant Ecology 182*, 209-233.
[78] Kuypers, M.M.M.; Blokker, P.; Hopmans, E.C.; et al. (2002). *Palaeogeography, Palaeoclimatology, Palaeoecology 185(1-2)*, 211-234.
[79] Reiss, C.; Blanc, P.; Trendel, J.M.; et al. (1997). *Tetrahedron 53(16)*, 5767-5774.
[80] Kirbii, A.; Lemée, L.; Chaouch, A.; et al. (2001). *Fuel 80*, 681-691.
[81] Amblès, A.; Grasset, L.; Dupas, G.; et al. (1996). *Org. Geochem. 24*, 681-690.
[82] Richnow, H.H.; Jenisch, A. & Michaelis, W. (1991). *Org. Geochem. 19(4-6)*, 351-370.
[83] Dragojlovic, V.; Amblès, A. & Vitorovic, D. (1993). *Journal of the Serbian Chemical Society 58*, 25-38.
[84] Dragojlovic, V.; Bajc, S.; Amblès, A.; et al. unpublished results.
[85] Dragojlovic, V.; Bajc, S.; Amblès, A.; et al. (2005). *Org. Geochem. 36(1)*, 1-12.
[86] Dragojlovic, V. unpublished results.
[87] Yoshioka, H. & Ishiwatari, R. (2005). *Org. Geochem. 36(1)*, 83-94.
[88] Peng, P.; Morales-Izquierdo, A.; Lown, E.M.; et al. (1999). *Energy and Fuels 13*, 248-265.
[89] Li, C.; Peng, P.; Sheng, G.; et al. *Org. Geochem. 35*, 531-541.

Chapter 5

MUCOUS MACROAGGREGATES IN THE NORTHERN ADRIATIC

Nives Kovač,[a] Jadran Faganeli[a] and Oliver Bajt[a]
[a]Marine Biology Station, National Institute of Biology, Fornače 41,
SI-6330 Piran, Slovenia

ABSTRACT

The episodic hyperproduction of mucous macroaggregates in the northern Adriatic, offers a rare opportunity to study the assembling of macromolecular DOM into macrogels and macroaggregates. They represent an important site of accumulation, transformation and degradation of organic matter, contributing to the patchiness, distribution and fate of particulate matter in seawater. In this chapter the results of our research work of several years is presented and combined with the results of some other authors. To elucidate this phenomenon, their biological and chemical composition, formation and degradation processes and finally their environmental role are presented and discussed. Emphasis is given to the use of spectroscopic techniques, e.g. ^{1}H NMR, ^{13}C NMR and FTIR, which are usually used for the determination of the chemical composition of organic compounds, to decoding the macroaggregate composition and structure.

INTRODUCTION

Mucillaginous material is important in many marine flux processes (Decho, 1990). Hyperproduction of mucous macroaggregates in the northern Adriatic (Figure 1) has been observed for more than 275 years (Fonda Umani et al., 1989) and recently occurred during the summers of 1988, 1989, 1991, 1997, 2000, 2001, 2002, 2003 and 2004. These events have been the object of various biological, ecological and chemical (Vollenweider and Rinaldi, 1995; Funari et al., 1999; Giani et al., 2005a; Manganelli and Funari, 2003; Tomasino, 1996; Thornton, 1999; Sellner and Fonda Umani, 1999) studies. These macroaggregates are found, as different types, at the surface layer (Figure 2), in water-column (Figure 3), and as sedimented aggregates (Figure 4). The different types of macroaggregates are classified as small flocs, macroflocs, stringers, tapes, clouds, creamy and gelatinous surface layers (Stachowitsch et al., 1990), or flocs, macroflocs, stringers, ribbons, cobwebs, clouds, false bottom, blankets, creamy surface and gelatinous surface layers (Precali et al., 2005). Besides the local hydrodynamics (turbulence and shear), the proximity of the sea surface and density interfaces in the water-column also contribute to the shape and size of northern Adriatic aggregates (Žutić and Svetličić, 2000). Mucilage normally appears in late spring/early summer. The horizontal and vertical distribution and accumulation of macroaggregates is heterogeneous and time-dependent on different factors such as the conditions in the water-column, the size, form and composition of the aggregates (biological and chemical) and the environmental conditions (Herndl, 1992; Mingazzini and Thake, 1995; Degobbis et al., 1995; Stachowitsch et al., 1990; Rinaldi et al., 1995). The buoyancy, derived from gas bubble formation, and density of aggregates also contribute to their vertical distribution (Herndl et al., 1992).

Similar events are also known for the continental coastal waters of the North Sea. Sometimes during spring, the French, Belgian and Dutch beaches are covered with brownish and stinking foams (Lancelot, 1995; Hamm and Rousseau, 2003). Accumulation of mucilage material was also observed around the Dalmatian coast (Stachowitsch et al., 1990), in Greek waters (Gotsis-Skretas, 1995), along Sicily (Calvo et al., 1995) and Sardinia (Olianas et al., 1996), in the Tyrrhenian Sea (Innamorati, 1995; Rinaldi et al., 1995; Mecozzi et al., 2001; Giuliani et al., 2005) and the Black Sea (Moncheva; pers. comm..), but on such a massive scale the phenomenon seems to be unique for northern Adriatic. Northern Adriatic can be therefore considered as a "natural laboratory" for the study of marine mucous macroaggregates.

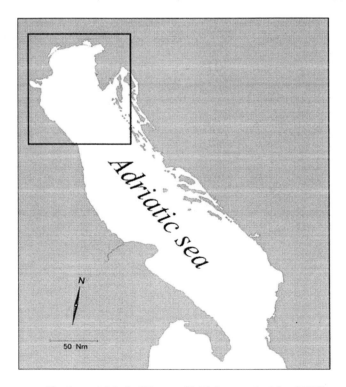

Figure 1: Sampling area – Northern Adriatic (Picture: T. Makovec, Archive MBS).

Figure 2: Creamy surface layer covering large areas of the northern Adriatic (Photo: N. Kovac).

Figure 3: Water-column aggregates in the form of clouds (Photo: T. Makovec, Archive MBS)

Figure 4: Sedimented macroaggregates negatively influence the benthic organisms (Photo: T. Makovec, Archive MBS).

COMPOSITION

Biological Composition

Northern Adriatic macroaggregates are characterized by a heterogenous composition, comprising phytoplankton, bacteria and cyanobacteria, mesozooplankton and microzooplankton and zooplankton debris (i.e. crustacean cuticles and antennae, faecal pellets), yeasts, pollen and various inorganic components such as the empty frustulaes of diatoms and skeletal remains of coccolithophorids, empty thecae of dinoflagellates and mineral particles (Figure 5). Diatoms (Figure 6) are usually reported as the dominant group of macroaggregates (Fanuko and Turk, 1990; Stachowitsch et al., 1990; Revelante and Gilmartin, 1991; Herndl, 1992; Degobbis et al., 1995; Baldi et al., 1997; Najdek et al., 2002; Kovač et al., 2005) but also other groups, such as dinoflagellates, microflagellates and coccolithophorids are present. Among the microflagellates, organisms belonging to the classes of euglenophytes and raphidophytes were identified (Kovač et al., 2005).

The characteristic/typical species present in macroaggregates were: a) diatoms: *Cylindrotheca closterium*, *Cyclotella* sp., *Pseudo-nitzschia pseudodelicatissima*, *Sceletonema costatum*, *Chaetoceros* sp., *Cerataulina pelagica*, *Thalassiosira* sp., *Leptocylindrus danicus*, *Rhizosolenia alata*,; b) dinoflagellates: *Prorocentrum* species such as *P. triestinum*, *P. minimum*, *P. micans*, *P. gracile*, *Heterocapsa* sp., *Ceratium furca*, a naked *Gymnodinium*-like dinoflagellate; c) coccolithophorids: *Calyptrosphaera oblonga*, *Emiliania huxleyi* and *Syracosphaera pulchra* (Marchetti et al., 1989; Pettine et al., 1993; Baldi et al., 1997; Penna et al., 1999; Kovač et al., 2005; Totti et al., 2005). However, the dinoflagellate *Gonyaulax fragilis* (Pompei et al., 2003; Del Negro et al., 2005; Mozetič, pers. comm.) could populate the macroaggregates. The mesozooplankton species composition (polychaete larvae, juvenile turbellarians, harpacticoid copepods, *Temora stylifera*) in the aggregates and marine snow of the northern Adriatic have been described by Bochdansky and Herndl (1992).

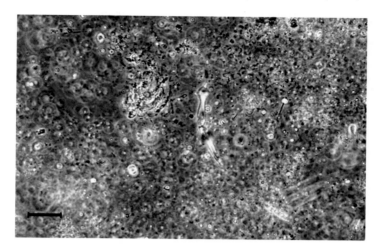

Figure 5: Light microscopy of heterogeneous composition of dense and gelatinous material of surface macroaggregate (Photo: P. Mozetič; reproduced from Kovač et al., 2005).

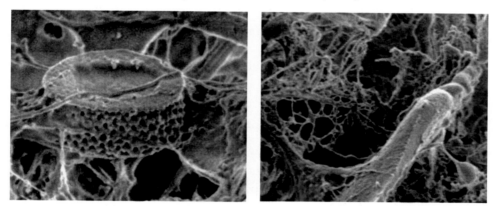

Figure 6: Cryo-SEM of macroaggregates: (a) A diatom embedded in the organic frame; (b) A view of the thin organic network and of a thick wall; (Courtesy of J. Trichet and C. Défarge, Institut des Sciences de la Terre d'Orléans, UMR 6113 CNRS-Université d'Orléans, France).

Chemical Composition

Elemental Composition

Macroaggregates from the gelatinous surface layer contained more than 90% water in accordance with their gel-like nature (Alldredge and Crocker, 1995). Organic matter varied between 35-57 % with an organic matter:organic carbon ratio in the range 2 and 3.6 (Posedel and Faganeli, 1991) which is similar to values reported by Giani et al. (2005b) and Pettine et al. (1995). The organic C and total N contents varied from 5-35% and 0.5-4.4%, respectively (Table 1). The C_{org}/N atomic ratios in mucous macroaggregates ranged from 5.8 to 28.7. These C/N ratios are higher than the classical Redfield ratio and fall within the range of C/N ratios observed from DOC and DON analyses, as well as UDOM samples (McCarthy et al., 1993; Biersmith and Benner, 1998), but differ greatly from typical C/N ratios of 35-45 observed for marine humic isolates (Meyers-Schulte and Hedges, 1986). The rather high C/N atomic ratios suggest a complex composition of mucous macroaggregates with a low protein content and the presence of degradation products. The C and N stable isotope composition of macroaggregates showed $\delta^{13}C$ and $\delta^{15}N$ values of about -17 to -21.3 ‰ and 5 ‰, respectively, indicating their phytoplanktonic origin (Posedel and Faganeli, 1991; Faganeli et al., 1995).

The inorganic component is mostly a result of scavenging and incorporation of autohtonic and allochtonic particles (Kovač et al., 2005). It usually ranged from 20 to 80% and was lower in aggregates sampled at the beginning of mucilage events, but usually higher in macroaggregates from the bottom layer and in sedimented samples (Giani et al., 2005b; Kovač et al., 2005; 2006). The higher relative abundance of mineral components (such as quartz and calcite) in the sedimented macroaggregates is due to the degradation processes and the contribution of sediment resuspension (Kovač et al., 2005; 2006).

Table 1: Literature data for the elemental composition (C_{org}, N, C/N) of mucous macroaggregates from the northern Adriatic.

Author	C_{org} (%)	N (%)	C_{org}/N
Posedel and Faganeli, 1991	10 – 25	0.7 – 3.4	8.6 – 28.7
Pettine et al., 1993	10.3 – 28.0	1.1 – 4.4	6.8 – 14.4
Pettine et al., 1995	10.3 – 35.2	1.1 – 4.4	5.8 – 12.6
Mecozzi et al., 2001	14.8, 28.8	1.4, 2.2	
Kovač et al., 2002	7.7 – 34.7	0.6 – 2.0	11.1 – 22.5
Giani et al., 2005b	11.1 – 32.3	0.7 – 4.3	9 – 16

There are only a few reports giving the more detailed elemental composition of mucilage and until now no detailed studies of interactions of mucous macroaggreagates with trace elements have been published. Elemental analysis of surface and water-column macroaggregates, sampled in the Northern Adriatic sea in July of 1991, showed the presence of major inorganic elements such as P, Si, Al, Fe, Mn, Ca, Mg, Na, K, Ti (Pettine et al., 1993; 1995). The mean percentage values of S, Si, Al and Fe were 1.1 ± 0.5, 6.4 ± 4.5, 1.5 ± 1.2 and $1.11 \pm 0.6\%$ (Pettine et al., 1995). The average values of Mn and Ti amounted 0.05 and 0.09% of dry macroaggregate mass (Pettine et al., 1993). The Fe, Al, Ti and Mn content are mostly influenced by mineral components (Pettine et al., 1993), while S and Si may result from biogenic and minerogenic sources (Pettine et al., 1995). In the surface macroaggregates of 1997 (Penna et al., 2000) similar elemental content was observed to those in samples of 1991 i.e. 0.65% Fe, 0.05% Mn and 5.49% Si. In those samples Cr, Pb and Ni were also identified, comprising 10, 14 and 9 ppm (Penna et al., 2000). The high iron and silica concentrations in mucilage were assigned to diatoms, which were the dominating group in the studied samples (Penna et al., 2000). EDS and X-ray analyses (Kovac et al., 2002) of the macroaggregates from July 1997, sampled in Piran Bay (Gulf of Trieste), showed a similar elemental composition, including elements of the major seawater ions, calcite, quartz and clay minerals i.e. Na, Cl, C, Al, Si, Mg, Ca, Fe and S. Besides Pb and Ni (123 and 9.9 ppm), Zn,

Cu, Cd and Co (2.3, 7.3, <0.3, <0.3 ppm) were also determined using polarographic methods (Mislej, pers. comm.). The mean Al and Fe content (1.21 ± 1.1% and 0.86 ± 0.6%) measured in different types of northern Adriatic macroaggregates sampled during the summers of 2000, 2001 and 2002 were comparable to previously mentioned values (Giani et al., 2005b). Higher values of Al and Fe were found in surface samples but Ca and Mg (slightly enriched) were more abundant in the sedimented samples. Giani et al. (2005b) also reported a high mean biogenic silicon content (29 ± 18%) indicating a significant presence of diatoms in aggregates.

Table 2: Comparison of trace metal concentrations in mucous macroaggregates ([1]Planinc, pers. comm.; [2]Dolenec, pers. comm.; [3]Falnoga, pers. comm.), suspended particulate matter ([1]Planinc, pers. comm., average values from 1983-90) and surficial sediment ([4]Faganeli et al., 1991) sampled in the southern part of the Gulf of Trieste (northern Adriatic).

Element	Macroaggregates (ppm, Fe %)			Suspended particulate matter[1] (ppm)		Surfical sediment[4] (ppm)
	July 1991[1]	June 2000[2]	June 2004[3]	Surface	Bottom	
Cd	0.04	0.1	< d.l.	4.32 ± 8.00	3.52 ± 5.48	0.23 ± 0.20
Cr	4.1	10.1	40.8	65.33 ± 84.81	75.50 ± 80.98	59.40 ± 43.4
Cu	3.14	19.8		48.65 ± 65.95	208.98 ± 280.53	20.80 ± 10.2
Zn	8.32	54.7	4.0	1802.46 ± 3058.29	1897.22 ± 548.76	101.90 ± 32.6
Hg	0.14	3.7	0.03	5.42 ± 7.07	6.70 ± 10.48	0.31 ± 0.24
Mn	108.6	257.6		294.25 ± 412.04	267.96 ± 278.33	592.8 ± 201.6
Ni	3.62	16.5		23.57 ± 35.56	64.91 ± 94.12	101.90 ± 63.7
Pb	1.25	10.9		47.99 ± 73.62	143.35 ± 156.61	18.90 ± 5.9
Fe	0.1	0.54	0.07	0.23 ± 0.40	0.36 ± 0.54	2.20 ± 0.6

Considering the metals in macroaggregates (Table 2), collected in the Gulf of Trieste (northern Adriatic), the following sequence was observed: Fe > Mn > Zn > Ni, Cu > Cr, Pb > Hg > Cd, that is not very different from Irving-Williams order of complex stability (Stumm and Morgan, 1996). The average concentration of bioactive metals (Fe, Mn, Zn, Ni, Cu) was >10 ppm, similarly observed also for colloidal isolates (Doucet et al., 2007), while that of Cr, Pb, Hg, Cd was <10 ppm. Comparison between metal content in macroaggregates and particulate matter (Table 2), sampled in the Gulf of Trieste, showed higher content in particulate matter and surficial sediment. This is probably due to various efficient carriers in particulate matter. Lower metal concentrations in macroaggregates could indicate a lack of binding sites in the organic and/or inorganic component. However, other possible interactions exist, including the entrapment of inorganic particles carrying metals. Carbonates and quartz, which represent an important inorganic constituent of macroaggregates, are thought not so important in trace element chemistry probably due to their lower complexing capability (Doucet et al., 2007). We also observed that sampling preparation (desalting), procedures and characteristics of specific mucilage event (terrestrial and aerosol inputs, duration of event, meteorological and hydrological conditions…) also contribute to the element concentrations. However, the biological composition (predominance of diatoms) and degradation stage of macroaggregates could also affect the content/release and transfer of heavy metals in seawater. The diatoms exhibit a strong affinity for trace elements, probably due to their organic coating that seems to be important in determining its amphoteric characteristics and the affinity to metals (Gélabert et al., 2004). The degradation of organic membrane alters the surface reactivity and metal binding properties of the diatom cell walls leading to release of trapped metals (Pokrovsky et al., 2002; Gélabert et al., 2004). Consequently, all above mentioned parameters should be considered in the future study of macroaggregates-trace elements interactions.

Molecular Composition

Gross chemical analyses of macroaggregates (Posedel and Faganeli, 1991; Penna et al., 2003) revealed carbohydrates to be the most abundant component (12-34%), followed by proteins (1-12%) and lipids (0.1-8%). Analyses of neutral monosaccharides showed that macroaggregates are composed of heteropolysaccharides, with glucose as the dominant neutral monosaccharide, followed by mannose, fructose, galactose, arabinose, ribose, xylose and fucose (Faganeli et al., 1995), suggesting their main origin from phytoplanktonic structural heteropolysaccharides (Haug and Myklestad, 1976; Hama and Handa, 1992). Glucose can also partially originate from more degradable water soluble reserve glucans (Handa, 1969; Handa and Yanagi, 1969), that is in accordance with the analysed content of water soluble carbohydrates (up to 2%; Posedel and Faganeli, 1991). This composition is similar to that of deep sea ultrafiltered dissolved organic matter (McCarthy et al., 1996). A somewhat different monosaccharide composition was reported by Giani et al. (2005b), consisting of galactose, glucose, mannose, xylose, rhamnose, fucose, ribose and arabinose. Differences could arise from compositional differences of macroaggregates in various developmental stages and due to selective degradation (Hama and Yanagi, 2001). Phytoplankton species, their nutrient condition, growth and physiological stage also affect the monosaccharide pool of macroaggregates (Pistocchi et al., 2005). The concentration of uronic acid residues is low (<5%; Giani et al., 2005b). Analyses of water soluble oligosaccharides, considering the mucilage precursors and degradation products, revealed a rather uniform composition comprising maltotriose, maltotetraose, maltopentaose, maltohexaose and maltoheptaose (Penna et al., 2003; Capiello et al., 2007). The dominating molecular class with avarage value of ≈287, determined in water-soluble fraction of mucous–macroaggregates by using size exclusion chromatography (SEC), also suggests the importance of degarded oligosaccharides (Figure 7). Additionaly, the presence of a higher molecular weight fraction was indicated by a relative molar mass of 2723 and 10^{10}. The amino acid composition (mol %) of macroaggregates, composed of (in decreasing order) glycine, aspartic acid, alanine, glutamic acid, leucine, valine, threonine, serine, isoleucine, lysine, proline, phenylalanine, arginine, histidine and tyrosine, is similar to summer POM in the northern Adriatic not affected by macroaggregate formation (Posedel and Faganeli, 1991). No sulphur aminoacids were detected. The mucilage lipids are mainly composed of triacylglycerols, hydrocarbons and free acids (Baldi et al., 1997). The macroaggregate fatty acids most probably originate from embedded phytoplankton (Najdek, 1996). The fatty acid ratios (16P/18P, 0.4-4.4), bacterial fatty acid proportions (3-6.7%) and ratios (C15:br/C15:0, 1.2-4.6) in fresh macroaggregates are lower and similar to northern Adriatic POM, than those determined in aged macroaggregates (16P/18P, 3.9-7.7; bacterial fatty acids 13.3-17.1%; C15:br/C15:0, 4.4-6.0) which resemble high photosynthetic activity of phytoplankton and bacteria (Najdek et al., 2002). The main source of the mucilage lipopolisaccharides is probably the degradation of cell membranes and the association of polysaccharides and lipids as storage materials (Baldi et al., 1997). Lipids from microzooplankton and micrometazoan captured from seawater

(Viviani et al., 1995) might also contribute to lipidic fraction. A high lipase and chitinase activity observed in mature aggregates could be due to accumulation of lipid droplets released by the zooplankton community (Müller-Niklas et al., 1994; Bochdansky and Herndl, 1992). Some lipopolysaccharides are the exudation products of bacteria (Decho 1990; Schnaitman and Klena, 1993).

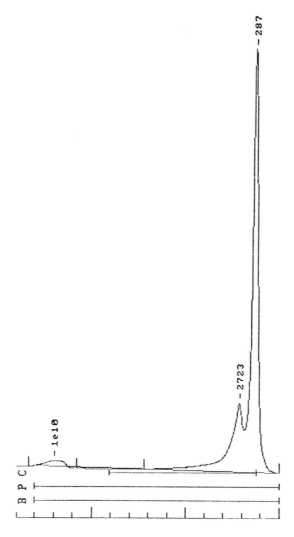

Figure 7: The high molecular weight of the water-soluble fraction of mucous macroaggregates was confirmed by the size exclusion chromatography (SEC).

Various spectroscopic techniques, which are usually used for the determination of the structure of organic compounds, were also applied in the case of mucous macroaggregates. Among them, NMR spectroscopy (^{1}H, ^{13}C) and FTIR spectroscopy seem to be the most useful and were already used when analysing aggregates from the mucilage events of 1988 (Marchetti et al., 1989) and 1991 (Kovac et al., 1995). After that many spectroscopic studies of the northern Adriatic macroaggregates have been performed.

In general, four major classes of structural elements can be identified from the ^{1}H-NMR spectra (Figure 8): carbohydrates (CHO) at $\delta=3.4-5.8$ ppm, aliphatic components (R) at $\delta=0.8-1.8$ ppm, functional groups such as ester and amide groups (COR) at $\delta=2.1-2.7$ ppm and organosilicon compounds from diatoms and silicon complexed to the organic matter of the macroaggregates (Si-R, Si-O-R) $\delta<0.7$ ppm (Kovac et al., 2002). Very small amounts of aromatic structures were detected. Considering the NMR evidence, it emerges that carbohydrates, polymethylene chains and carboxyl groups are present and the polymethylene chains compose a part of the bound carboxylic acids (Kovac et al., 1998). Similar compositional features were confirmed using ^{13}C NMR spectroscopy. The spectra (Figure 9) showed important signals in the region at $\delta=0-50$ ppm, assigned to paraffinic carbons (Poutanen, 1985; Kalinowski and Blondeau, 1988; Guggenberger et al., 1994; Wilson et al., 1981; Knicker et al., 1996). Strong signals in the range $\delta=65-100$ ppm which were assigned to the carbon atoms in carbohydrates (Benner et al., 1990; Gamble et al., 1994; McCarthy et al., 1993; de Beer et al., 1994; Sihombing et al., 1996; Guggenberger et al., 1994; Knicker et al., 1996). The resonances at 50-110 ppm could also be due to carbons bound to heteroatoms (Allard et al., 1997), most likely to oxygen. The resonances around $\delta=100$ ppm can be assigned to the anomeric C-1 carbon in carbohydrates (Benner et al., 1990; Guggenberger et al., 1994) and/or -O-C-O- functionalities (Poutanen, 1985). The signals between $\delta=110-160$ ppm are assignable to aromatic-C and olefinic-C which are not distinguishable (Wilson et al., 1981; Poutanen, 1985). From the solid state ^{13}C NMR spectrum of the bulk macroaggregate sample it emerged that the aromatic and olefinic structures comprise an important part of the macroaggregate composition, up to 15%. This difference is most probably due to the high degree of substitution of aromatic and olefinic C atoms. Solid state ^{13}C NMR spectroscopy also confirmed the presence of carbonyl, carboxyl and amide groups (160-220 ppm), comprising around 6% of the macroaggregate composition.

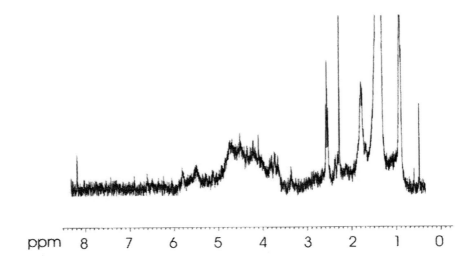

Figure 8: ^{1}H-NMR spectra of macroaggregates from the summer of 1991 (reproduced from Kovac et al., 1995).

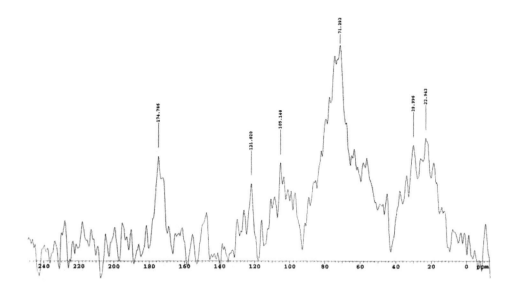

Figure 9: Solid-state ^{13}C-NMR spectrum of a surface macroaggregate sample collected in the Gulf of Trieste on June 13, 2000 (reproduced from Kovač et al., 2006).

FT-IR spectroscopy (Kovac et al., 2002; 2004; 2006) also used for determination of the structure of macroaggregates, confirmed the presence of same major components (Figure 10). A better insight into the inorganic part of the macroaggregate structure was also achieved. The spectra exhibit absorption bands in the range 2800-3600 cm^{-1}, corresponding to those of hydrogen bonded OH and NH groups (Zegouagh et al., 1999), and weaker absorption in the range 2800-3000 cm^{-1} originating from various aliphatic components. In the range 400-1800 cm^{-1}, numerous bands were noted and attributed to the vibrations of organic (proteins, polysaccharides, carbonylic and etheric structures) and inorganic components (carbonates and silicates). Despite the problem of overlapping of some characteristic bands, the band at 877 cm^{-1}, together with the bands at 712, 877, 1425-1434 cm^{-1}, can be assigned to calcite. The presence of silicates and organosilicon compounds (Anderson, 1974; Bellamy, 1975; Groselj et al., 1999) could be determined by the bands in the region from 900 to 1200 cm^{-1}, which coincide with the most characteristic bands of polysaccharides (~1100 cm^{-1}) (Guggenberger et al., 1994). Some weak bands below 800 cm^{-1} additionally confirm the presence of various silicate compounds. Similar major components in macroaggregate samples from the northern Adriatic were also determined by other authors (Mecozzi et al., 2001; 2004; 2005; Berto et al., 2005).

For better elucidation of the composition of macroaggregates, the bulk macroaggregates were separated into water soluble (of lower molecular weight) and water insoluble (more refractory, of higher molecular weight) fractions (Kovac et al., 2004). ^{1}H-NMR spectra after the partition of bulk macroaggregates into two fractions showed similar major components as in the case of the bulk macroaggregates. Comparison of the relative composition, according to the H integrals in the NMR spectra, revealed that the water-insoluble fraction of macroaggregates consists mainly of aliphatic structures, probably bonded to carbohydrates through ester and amide bonds, organosilicon compounds of diatomaceous frustulae and carbonates. The prevailing presence of mineral particles, especially calcite, quartz and silicates was also confirmed by XRD and EDS analyses (Kovac et al., 2002). On the other hand, the water-soluble fraction is mostly composed of carbohydrates with a minor amount of aliphatic component. The aliphatic nature of the water-insoluble fraction is also indicated by FT-IR analysis, i.e. by alkyl group bands at 2924 cm^{-1}, 2853 cm^{-1}, 1430 cm^{-1} and 1385 cm^{-1}. The FT-IR spectrum also evidenced a broad band centred at 3437 cm^{-1} indicating the presence of hydroxyl groups, as well as a large contribution of mineral component (calcite: 1429 cm^{-1}, 872 cm^{-1}, 710 cm^{-1}; silicates: 777, 527 and 470 cm^{-1}). The FT-IR spectrum of the water-soluble fraction showed a weaker absorbance (relative to the wide OH band at ≈3436 cm^{-1}) for the aliphatic components (2800-3000 cm^{-1}), suggesting the less lipidic nature of this fraction. It is characterized by a strong absorption centred around 3400 cm^{-1}, corresponding to the bands of OH and NH groups (Allard et al., 1997; Zegouagh et al., 1999). The presence of a unique nitrogenous component, most probably amides (band between 1560–1650 cm^{-1}), is also characteristic of the water soluble fraction, especially in its more developed stage. Amides probably originate from algal proteinaceous material and to a lesser extent from other microbial (bacterial) sources.

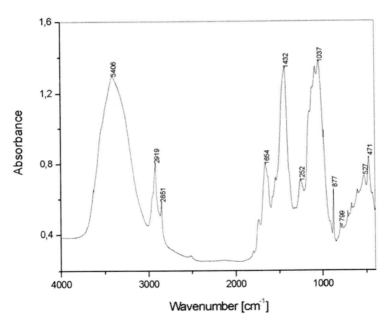

Figure 10: FT-IR spectra of macroaggregates (in KBr pellets, range 4000 - 400 cm^{-1}) collected in June 2004.

Partition of the bulk macroaggregate sample in two fractions revealed also some important data about the origin of mucous macroaggregates. Namely, the comparison of ^{1}H-NMR spectra of the water-soluble fraction of "fresh" aggregates collected at the beginning of the mucilage event and UDOM isolated from cultured diatom *C. fusiformis* revealed structural similarities with a higher content of polysaccharides (Kovac et al., 2004). The important compositional and structural similarities with the microalgal extracellular macromolecular DOM (Aluwihare and Repeta, 1999) and deep water UDOM (McCarthy et al., 1993) were also indicated (Kovac et al., 2002). This suggests that the UDOM might be the precursor of macroaggregate formation (see "Macroaggregate Formation").

Macroaggregate Interstitial Water

The interactions of macroaggregate matrix with different inorganic species has been also recognized from the higher salinity and lower pH of the interstitial water (24.94; 7.7), obtained after the centrifugation, comparing to values determined in the surrounding seawater (33.73; 8.1). Those results indicate the higher binding capacity of the mucilage matrix for alkaline ions Ca^{2+} or CO_3^{2-}. To the lower pH of interstitial water some oxidized degradation products (low molecular acids) and acidic free amino acids (Müller-Niklas et al., 1994) can also contribute. Dominant free amino acid of macroaggregates was glutamic acid followed by alanine, aspartic acid, leucine, arginine, serine, histidine, valine, isoleucine, glycine, threonine, tyrosine, phenylalanine, methionine. The mean concentrations of dissolved free amino acids of macroaggregates were different and higher than that in ambient water probably due to a high concentration of plankton (Müller-Niklas et al, 1994). Herndl (1992) found that macroaggregates from northern Adriatic were enriched in NO_2^-, NO_3^-, NH_4^+. Nutrient concentrations were orders of magnitude higher than in the surrounding water (Del Negro et al., 2005). On the contrary to the trend in ambient water, the PO_4^{3-} concentrations significantly increased in macroaggregates with time (Herndl, 1992).

MACROAGGREGATE FORMATION

While there is a wide consensus among researchers that mucilage is mostly composed of heteropolysaccharides produced by exudation (Myklestad, 1995) and cell lysis (Baldi et al., 1997), the formation of macroaggregates in the northern Adriatic is at present not completely understood. Most hypotheses (Degobbis et al., 1999) suggest the importance of channeling of primary production into the carbohydrate pool (Azam et al., 1999; Myklestad, 1995). The pelagic origin and formation of macroaggregates in the water-column was indicated by microscopic observations (Fanuko et al., 1989; Stachowitsch et al., 1990; Degobbis et al., 1995; Baldi et al., 1997; Mozetič, pers. comm.). Diatoms seem to be the principal producers of mucilage, although exudation and lysis of other organisms (see "Biological Composition") also contribute.

In the northern Adriatic, the months before a mucilage event (March – May) can be characterized as an "incubation" period with favourable conditions for organic matter production (Degobbis et al., 2005). The greater availability of free polysaccharides in late spring (Figure 10) seems to favour the formation of more complex chains and crosslinking producing gels (Chin et al., 1998) and macrogels of large dimensions (Verdugo et al., 2004). The aggregation process can be explained by polymer gel theory (Chin et al., 1998) through the formation of nanogels and later of microgels that continue to agglomerate into macrogels and particulate organic matter (Verdugo et al., 2004; Svetličić et al., 2005; Žutić et al., 2004). The transformation of macromolecular DOM to POM includes an increase in size and changes of reactivity of the material. Macromolecules represent a less reactive substrate for bacterial degradation and they can concentrate (Azam et al., 1999; Herndl et al., 1999; Mari et al., 2007). Degradation of organic macromolecules into molecules sufficiently small (approximately 600 Da; Weiss et al., 1991) to pass into the cell depends on the ability of some fraction of the microbial community to generate specific extracellular enzymes (Ziervogel et al., 2007). The uncoupling between phytoplanktonic and bacterial development could result from the macromolecular nature of the dissolved organic matter released from phytoplankton (Billen and Becquevevort, 1991). However, the hypothesis that extracellular enzymatic hydrolysis is a slow step in macromolecular organic matter breakdown in the sea is not always confirmed (Arnosti et al., 1994).

The origin of elevated quantities of dissolved carbohydrates should be related to the higher phytoplankton biomass occurring before the appearance of macroaggregates (Penna et al., 2004) since phytoplankton is known to be a major source of polysaccharides necessary for the formation of exopolymer particles (Logan et al., 1995; Ramaiah et al., 2001; Ahel et al., 2005) and subsequently mucilage (Radić et al., 2005). Concentrations of exopolysaccharide precursors in seawater often increase during phytoplankton blooms, especially when phytoplankton becomes nutrient-limited (Mari and Burd, 1998; Corzo et al., 2000; Engel et al. 2002; 2004). However, the relation between phytoplankton biomass, expressed as chlorophyll *a* concentrations, and concentrations of dissolved carbohydrates is at present unclear. Moreover, some recent studies indicate no significant increase in phytoplankton biomass in years with mucilage events (Fonda Umani et al., 2005; Precali et al., 2005). It appears from phytoplankton nutritional requirements that the presence of higher nutrient N and P concentrations alone is a less important factor than the change of limiting nutrient and DIN/PO_4^{3-} ratio (Cozzi et al., 2004). This can occur in the surface sea layer in conditions of marked retention of freshened waters in the northern Adriatic basin and water-column stratification (Degobbis et al., 2005). Nitrate and later phosphate seem to be intensively consumed before mucilage appearance (Figure 11), causing a variation of DIN/PO_4^{3-} ratios (Faganeli et al., 1995) which can produce nutritional stress to algae, leading to increased polysaccharide excretion with consequent formation of macroaggregates (Myklestad 1995; Maestrini et al., 1997; Staats et al., 2000). P-limitation may, on the other hand, limit the bacterial degradation of exudates (Azam et al., 1999) allowing their accumulation with successive aggregation. In the northern Adriatic, phosphorus is thought to be the primary biolimiting element, before and after the substantial reduction of the phosphorus load of the Po River (Degobbis et al., 2005).

The summer hydrographic isolation of the northern Adriatic Sea (Supić and Orlić, 2000) with its specific oceanographic conditions, including the formation of gyre, higher seawater residence time, and development of a pronounced pycnocline in the stable summer conditions with low turbulent shear additionally enables the subsequent concentration and agglomeration of macromolecular organic matter and phytoplankton cells. The organic-mineral associations probably also have an important role in the formation, evolution and transformation of the mucilage phenomenon. Mineral particles and ions such as Ca^{2+} and Fe^{3+} seem to be efficient cross-linkers (Verdugo 1994; Chin et al., 1998; Thornton et al., 1999). Accordingly to the tendency for dissolved organic matter to concentrate at phase boundaries, such as the sediment-water and air-sea interfaces, and the pycnocline, and due to the presence of mineral particles, colloidal (macromolecular) organic matter can agglomerate and macroaggregates further accumulate, especially in these layers. The increased assembly rates of marine gels leading to larger particles contribute to clarification of the sea water-column, a phenomenon which was usually observed just before the appearance of greater macroscopic mucous event. In this case macroaggregates play an important role in scavenging processes and further in the transformation of dissolved marine organic matter. Therefore, in addition to higher polysaccharide concentration levels, the presence of particles (Kovac et al., 2002; 2004), the above mentioned specific hydrological conditions including temperature, stratification, reduction of water exchange with the central Adriatic and low turbulence (Ahel et al., 2005), photochemical reactions including photooxidation and photopolymerization (Kovac et al., 1998), and the quality and assemblage of dissolved organic matter (Mecozzi et al., 2004; Žutić et al., 2004) and exudates can all be important factors for mucilage formation and evolution in the northern Adriatic.

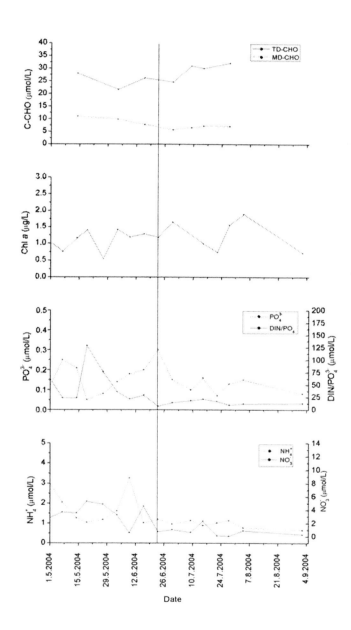

Figure 11: Variations of total dissolved carbohydrates and dissolved monosaccharides (Penna, pers. comm.), chlorophyll *a* (Mozetič, pers. comm.), nitrate, ammonium, dissolved inorganic nitrogen (DIN), phosphate and DIN/PO_4^{3-} ratio in the Gulf of Trieste. Vertical line indicates the occurrence of macroaggregates.

DEGRADATION AND PERSISTENCE OF MACROAGGREGATES

Bacteria seem to be the primary degraders of mucilage (Minganzzini and Thake, 1995), since with maturation and ageing their production increases (Müller-Niklas et al., 1994). Macroaggregate biopolymers are also subjected to various chemical transformation including degradation-recondensation and condensation-polymerization with other low molecular weight compounds, as well as photochemical transformations. Study of the temporal variations of the relative composition of macroaggregate samples (Kovac et al., 2004) showed nearly equal percentages of aliphatic structures and carbohydrates in the fresh macroaggregates. Their maturation leads to a faster decrease of the relative content of carbohydrates in comparison to the aliphatic component, and an increase of the relative content of organosilicon compounds. This is an indication of cleavage of the glycosidic linkage as the initial and the most important degradation process. The presence of β- and α-glycosidic linkages is evident from our extensive spectroscopic studies. We also observed the connection between the phytoplankton composition and enzymatic activity. The species composition has been reflected in the more pronounced degradation of α-glycosidic bond compared to β-glycosidic linkage, i.e. in the sample from summer 2004 (Mohar et al., 2007) with an important part of species (*Gonyaulax fragilis*) containing α(1-4) glucans as reserve polysaccharide. A shift of α-glycosidic bonds to β-glycosidic linkages in more refractory polysaccharide material was reported for aged aggregates (Herndl,
1992; Muller-Niklas et al., 1994) that are usually dominated by diatoms which contained β-glucan.

The preservation of lipidic fraction was also suggested by the study of structural transformations of mucous macroaggregates after the enzymatic hydrolysis (Mohar et al., 2007; in prep.) using proteinase, pronase, amylase, glucosidase and lipase. FTIR analyses of macroaggregates before and after the addition of enzymes confirmed the existence of important associations between lipids, carbohydrates and minerals and weaker interactions with proteins. Those results indicate that glycolipids from bacterial and phytoplankton cells might be the most important components resistant towards enzymatic hydrolisis. Besides this, the preservation of a part of the labile organic nitrogen incorporated into refractory organic matter was observed. A higher nitrogen content and lower C_{org}/N ratio of the more developed/matured sample in comparison to the values were determined in sample from the beginning of a mucilage event (Kovac et al., 2004). This emerged also from the ^{13}C NMR spectra (Kovač et al., 2006) showing an increase of signals in the region from 45 to 60 ppm, most probably assigned to N-alkyl C atoms. This proteinaceous material could be stabilized and preserved through encapsulation inside macromolecular matrix (Nguyen and Harvey, 2003; Zang and Hatcher, 2002; Knicker and Hatcher, 1997) or/and sorptive protection on minerals (Mayer, 1999; Keil et al., 1994). According to the importance of diatoms in mucilage, also refractory silicalemma proteins of diatom cell walls were proposed to contribute to preserved N-organic component (Kovač et al., 2006).

Photochemical degradation is also an important degradation process of organic matter in seawater. Despite the very complex and rigid structure of mucous macroaggregates with no important chromophoric constituents, photochemical degradation can be promoted by many photosensitizing constituents of seawater, i.e. humic substances, inorganic ions. Photochemical degradation of the photo-labile water soluble fraction of macroaggregates proceeds by cleavage of glycosidic linkages and later by degradation of the monosaccharides produced (Kovac et al., 1998). After cleavage of anomeric linkages, photorepolymerization may also occur, leading to high molecular weight organic matter. Our results show that photochemical degradation proceeds in parallel with microbial degradation. The lower molecular weight photoproducts could be subsequently utilized by microbes as substrates for microbial growth. In this way these two degradation processes may simultaneously lead to the more efficient degradation of macroaggregates.

Considering the general structure of macroaggregate organic matter, mainly composed of carbohydrates, lipids and proteins with a high microbial density and photochemical transformation, this organic matrix represents quite a stable marine assemblage resistant to a breakup. The duration of mucilage events could continue from one to several months, i.e. from late May/June to September. The complex structure and stability of macroaggregates via mineral-organic matter interactions (Kovac et al., 2002) influence the persistence/resistance of macroaggregates in the summer stratified northern Adriatic water-column, when production and sedimentation of macroaggregates are hindered (Posedel and Faganeli, 1991; Faganeli et al., 1995). Additionally, hydration of polysaccharides and consolidation with metals and/or lipids contribute to the resistance of the mucilage to microbial attack (Baldi et al., 1997). Ciglenečki et al. (2000) proposed the reduced sulphur species as stabilizing agents of polysaccharides by using electrochemical methods and Raman spectroscopy. The disappearance of mucous macroaggregates is usually connected with rain storm events, alteration of the hydrographic structure of the water-column and modification of the water motion/circulation pattern (Grilli et al., 2005; Russo et al., 2005). Enhanced wind-induced turbulent mixing contributes to the break-up of the pycnocline and of macroaggregates resulting in an increase of sedimentation rate. The more abundant benthic microbial community (Herndl et al., 1987) probably contributes to more successful degradation in the late summer (Baldi et al., 1997). Grazing of aggregates by fish and zooplankton also proceed but on the other hand the copepod feeding could be inhibited by high molecular weight carbohydrate exudates (Malej and Harris, 1993). This reduced grazing pressure is thought to be more important for mucilage development.

Investigations of species/particle composition and particle-organic matrix interactions have an important role in understanding the degradation processes and description of maturation stages of mucous aggregates. Electron microscopic studies of this highly hydrated organic netted material confirmed previous findings on marine snow (Decho, 1990; Heissenberger et al., 1996; Leppard et al., 1996) that polysaccharide fibrils constitute the structural framework of the mucous matrix (Giani, 2002; Kovač et al., 2005). The fibrils could be arranged in a more or less porous structure. In the mucous gel material, a repeating network of cavities between polymer strands has also been observed using atomic force microscopy (Svetličić et al., 2005). The more complex and rigid organization of the basic organic network, characterized by a tighter organization of the fibrils with "wall pieces" ("re-inforced walls"), including an abundant non-living component such as remains of broken cells, crustacean cuticles, detritus and several mineral particles, indicates a longer physical diagenesis of the polysaccharidic fibrils (Kovač et al., 2005) i.e. a mature, more developed or "older" stage of macroaggregates. The physiological stage of the phytoplankton cells (mostly senescent and degraded, and empty skeletal parts) with the predominance of a live diatom species – *Cylindrotheca closterium*, also suggests mature, more developed macroaggregates (Kovač et al., 2005). Similarly, Najdek et al. (2002) observed that most of aged large macroaggregates from summer 1997 were also strongly inhabited by *Cylindrotheca closterium* but its presence in such aggregates could be in some cases negligible. The pigment composition of the mucilage samples showed an increased diatom contribution to the total biomass, and a relative decrease of other phytoplankton with mucilage maturation/ageing (Flander-Putrle, 2003). The species often differ from those in the water-column indicating that mature macroaggregates represent self sustaining communities (Najdek, 1996) with phytoplankton-bacterial interactions that contribute to their residence time in water-column (Najdek et al., 2002).

ENVIRONMENTAL ROLE OF MACROAGGREGATES

Despite their irregular occurrence, mucilage events exert great influence on the marine environment, including natural processes, tourism, mariculture, fisheries and the economy of coastal countries.

Mucous aggregates attach themselves to fishing nets, aquaculture cages, influence the eggs/larval/juvenile development of marine organisms and make fishing difficult thus decreasing fish catches or production. Hyperproduction, accumulation and degradation of mucilaginous material negatively influence touristic activities, as well as decrease the aesthetic values of the marine environment and the quality of bathing water. Concerning the general composition of mucous macroaggregates, there is no risk for human health. However, some toxic impact (indirect effects) could be connected with the ability of macroaggregates to adsorb components such as heavy metals, algal toxins, pathogens etc. (Volterra, 1995; Funari and Ade, 1999).

Scavenging and accumulation of various particles and chemicals from seawater contribute to the decrease in concentration of these substances in ambient seawater. During mucilage events a significantly lower content of suspended matter and plankton biomass in the aggregate-free seawater is usually observed (Malej et al., 2001). In this way, suspended and sinking macroaggregates impact the whole plankton community (Fonda Umani et al., 2005). They affect the temporal and spatial variability, feeding capability and food web structure and function of microzooplankton and mesozooplankton (Kršinić, 1995; Bochdansky and Herndl, 1992; Malej and Harris, 1993, Cabrini et al., 1992; Milani and Fonda Umani, 1992; Cataletto et al., 1996; Fonda Umani et al., 2005). A great impact is also observed on benthos (Figure 4) after macroaggregate sedimentation. This can lead to the marked oxygen consumption and asphyxiation of sedentary organisms or/and anoxic conditions of affected area at the sea bottom. However, Cornello et al. (2005) reported that the soft-bottom macro-zoobenthos seems not to be directly affected by mucilage events.

The comparison of annual gross POM deposition in the Gulf of Trieste (northern Adriatic) in 1991 (characterized by the mucilage event) and 1992 (lack of mucilage) suggest that the macroaggregate outbreaks have minimal impact on annual organic C and N budgets (Kemp et al., 1999). Besides their important role in summer vertical flux and horizontal transport of accumulated constituents, macroaggregates represent an important site of abiotic and biotic transformations of mucilage material. This heterogeneous cross-linked organic matrix variously combined with inorganic component of macroaggregates, also offers a large

Conclusion

The process of mucous macroaggregates formation is important in understanding the cycling of organic matter in the sea. Mucilage production in the northern Adriatic is primarily associated with phytoplankton, mostly with diatom blooms. Macroaggregates are most probably the result of the agglomeration of macromolecular dissolved organic matter into macrogels and particulate organic matter under favourable environmental conditions. They are characterised by a heterogenic composition including phytoplankton, bacteria, mineral particles and skeletal remains. Their chemical composition is complex. Carbohydrates are the most abundant component followed by lipids and proteins. Applying different spectroscopic techniques four major classes of structural elements were identified: carbohydrates, aliphatic components, functional groups such as ester and amide groups and an important inorganic part (carbonates, silicates). The binding capacity of macroaggregates allows them to participate in scavenging processes and to influence the trace element chemistry in seawater. They are subjected to various microbial and photochemical transformations. Maturation of macroaggregates leads to the faster degradation of carbohydrates in comparison to the aliphatic component and a relative increase of the organosilicon compounds. The photochemical degradation of the photo-labile water soluble fraction of macroaggregates proceeds by cleavage of glycoside linkages and later by degradation of the produced monosaccharides. The macroaggregate phenomenon has an important impact on marine ecosystem (impact on plankton and benthos) and local economy (tourism, fisheries, mariculture). Although an extensive work has been performed, additional research work is still needed, especially the long-term measurements of relevant parameters on adequate time scale for better forecasting of mucilage events.

References

Ahel, M.; Tepic, N. & Terzic, S. (2005). *Sci Total Environ.* 353, 139-150.
Allard, B.; Templier, J. & Largeau, C. (1997). *Org Geochem.* 26, 691-703.
Alldredge, A.L. & Crocker, K.M. (1995). *Sci Total Environ.* 165, 15-22.
Aluwihare, L.I. & Repeta, D.J. (1999). *Mar Ecol Prog Ser.* 186, 105-117.
Anderson, D.R. (1974). In *Analysis of Silicones*; Smith, A.L.: (Ed.); Wiley-Interscience: NewYork, US, pp 264-286.
Arnosti, C.; Repeta, D.J. & Blough, N.J. (1994). *Geochim Cosmochim Acta* 58, 2639-2652.
Azam, F.; Fonda Umani, S.; Funari, E. (1999). *Ann Ist Sup Sanita* 35, 411-419.
Baldi, F.; Minacci, A.; Sailot, A.; et al. (1997). *Mar Ecol Prog Ser.* 153, 45-57.
Bellamy, L.J. (1975). *The Infra-red Spectra of Complex Molecules*; Chapman and Hall: London, UK, 433 pp.
Benner, R.; Hatcher, P.G. & Hedges, J.I. (1990). *Geochim Cosmochim Acta* 54, 2003-2013.
Berto, D.; Giani, M.; Taddei, P.; et al. (2005). *Sci Total Environ.* 353, 247-257.

Biersmith, A. & Benner, R. (1998). *Mar Chem.* 63, 131-144.
Billen, G. & Becquevort, S. (1991). *Polar Research* 10, 245-254.
Bochdansky, A. & Herndl, G. (1992). *Mar Ecol Prog Ser.* 87, 135-146.
Cabrini, M.; Fonda Umani, S. & Honsell, G. (1992). In *Marine coastal eutrophication*; Vollenweider, R.A.; Marchetti, R.; Viviani, R.: (Eds.); Elsevier: Amsterdam, NL, pp.557-568.
Calvo, S., Barone, R. & Naselli Flores, L. (1995). *Sci Total Environ.* 165, 23-31.
Cappiello, A.; Trufelli, H.; Famiglini, G.; et al. (2007). *Water Res.* 41, 2911-2920.
Cataletto, B.; Feoli, E.; Umani, S.F.; et al. (1996). *PSZN: Mar Ecol.* 17, 291-308.
Chin, W.C.; Orellana, M.V. & Verdugo, P. (1998). *Nature* 391, 568-57.
Ciglenečki, I.; Ćosović, B.; Vojvodić, V.; et al. (2000). *Mar Chem.* 71, 233-249.
Cornello, M.; Boscolo, R. & Giovanardi, O. (2005). *Sci Total Environ.* 353, 329-339.
Corzo, A.; Morillo, J.A. & Rodriguez, S. (2000). *Aquat Microb Ecol.* 23, 63-72.
Cozzi, S.; Ivančić, I.; Catalano, G.; et al. (2004). *J Marine Syst.* 50, 223-241.
de Beer, T.; van Zuylen, C.W.E.M.; Hard, K.; et al. (1994). *FEBS Lett.* 348, 1-6.
Decho, A.W. (1990). *Oceanogr Mar Biol Annu Rev.* 28, 73-153.
Degobbis, D.; Fonda Umani, S.; Franco, P.; et al. (1995). *Sci Total Environ.* 165, 43-58.
Degobbis, D.; Malej, A. & Fonda Umani, S. (1999). *Ann Ist Sup Sanita* 35, 373-381.
Degobbis, D.; Precali, R.; Ferrari, C.R.; et al. (2005). *Sci Total Environ.* 353, 103-114.
Del Negro, P.; Crevatin, E.; Larato, C.; et al. (2005). *Sci Total Environ.* 2005, 353, 258-269.
Doucet, J.F.; Lead, J.R. & Santschi, P.H. (2007). In: *Environmental Colloids and Particles. Behaviour, Separation and Characterisation;* Wilkinson, K.J.; Lead J.R.: (Eds.); IUPAC Series on Analytical and Physical Chemistry of Environmental Systems; John Wiley&Sons: New York, US, Vol. 10, pp. 95-157.
Engel, A.; Goldthwait, G.; Passow, U.; et al. (2002). *Limnol Oceanogr.* 47, 753-761.
Engel, A.; Thoms, S.; Riebesell, U.; et al. (2004). *Nature* 428, 929-932.
Faganeli, J.; Planinc, R.; Pezdič, J.; et al. (1991). *Mar Geol.* 99, 93-108.
Faganeli, J.; Kovac, N.; Leskovsek, H.; et al. (1995). *Biogeochemistry* 29, 71-88.
Fanuko, N.; Rode, J. & Drašlar, K. (1989). *Biol Vestn.* 4, 27-34.
Fanuko, N. & Turk, V. (1990). *Bolettino di Oceanologia Teorica ed Applicata* 8, 3-11.
Flander-Putrle, V. Ph. D. thesis (in Slovenian) (2003). University of Ljubljana, Biotechnical Faculty, Department of Biology; 158 pp.
Fonda Umani, S.; Ghiarardelli, E. & Specchi, M. (1989). Regione autonoma Friuli-Venezia Giulia, Trieste; 178 pp.
Fonda Umani, S.; Milani, L.; Borme, D.; et al. (2005). *Sci Total Environ.* 353, 218-231.
Funari, E.; Azam, F.; Fonda Umani, S.; et al. (Eds.) (1999). State of the art and new scientific hypotheses on the phenomenon of mucilages in the Adriatic Sea; *Ann Ist Sup Sanita*; Vol. 35, pp 353-426.
Funari, E. & Ade, P. (1999). *Ann Ist Sup Sanita* 35, 421-425.
Gamble, G.R.; Sethuraman, A.; Akin, D.E.; et al. (1994). *Appl Environ Microb.* 60, 3138-3144.
Gélabert, A.; Pokrovsky, O.S.; Schott, J.; et al. (2004). *Geochim Cosmochim Acta* 68, 4039-4058.
Giani, M. (2002). *Archo Oceanogr Limnol.* 23, 29-41.
Giani, M.; Degobbis, D. & Rinaldi, A. (Eds.) (2005a). Mucilages in the Adriatic and Tyrrhenian Seas; Sci Total Environ; Elsevier: Amsterdam, NL, Vol. 353, pp 1-379.

Giani, M.; Berto, D.; Zangrando, V.; et al. (2005b). *Sci Total Environ.* 353, 232-246.
Giuliani, S.; Virno Lambreti, C.; Sonni, C.; et al. (2005). *Sci Total Environ.* 353, 340-349.
Gotsis-Skretas, O. (1995). *Sci Total Environ.* 165, 229-230.
Grilli, F.; Paschini, E.; Precali, R.; et al. (2005). *Sci Total Environ.* 353, 57-67.
Groselj, N.; Gaberscek, M.; Opara Krasovec, U.; et al. (1999). *Solid State Ionics* 125, 125-133.
Guggenberger, G.; Zech, W. & Schulten, H. (1994). *Org Geochem.* 21, 51-66.
Hama, J. & Handa, N. (1992). *J Exp Mar Biol Ecol.* 162, 159-176.
Hama, T. & Yanagi, K. (2001). *Limnol Oceanogr.* 46, 1945-1955.
Hamm, C.E. & Rousseau, V.J. (2003). *Sea Res.* 50, 271-283.
Handa, N. (1969). *Mar Biol.* 4, 208-214.
Handa, N. & Yanagi, K. (1969). *Mar Biol.* 4, 197-207.
Haug, A. & Myklestad, S. (1976). *Mar Biol.* 11, 15-26.
Heissenberger, A.; Leppard, G.G. & Herndl, G.J. (1996). *Mar Ecol Prog Ser.* 135, 299-308.
Herndl, G.J.; Faganeli, J.; Fanuko, N.; et al. (1987). *PSZN I: Mar Ecol.* 8, 221-236.
Herndl, G.J. (1992). *Mar Microb Food Webs* 6, 149-172.
Herndl, G.J.; Karner, M.; Peduzzi, I.M.E. (1992). In *Marine Coastal Eutrophycation*; Vollenweider, R.A.; Marchetti, A.; Viviani, R.: (Eds.); Sci Total Environ (suppl.); Elsevier Science, Amsterdam, 525-538.
Herndl, G.J.; Arrieta, J.M. & Stoderegger, K. (1999). *PSZN I: Mar Ecol.* 35, 405-409.
Innamorati, M. (1995). *Sci Total Environ.* 165, 65-81.
Kalinowski, E. & Blondeau, R. (1988). *Mar Chem.* 24, 29-37.
Keil, R.G.; Montlucon, D.B.; Prahl, F.G.; et al. (1994). *Nature* 379, 549-552.
Kemp, W.M.; Faganeli, J.; Puskaric, S.; et al. (1999). In *Ecosystems at the land-sea margin: Drainage basin to coastal sea*. Malone, T.; Malej, A.; Harding, L.W.Jr.; Smodlaka, N.; Turner R.: (Eds); Amer. Geophys. Union Publ., Washington, DC, pp 295-339.
Knicker, H.; Scaroni, A.W. & Hatcher, P.G. (1996). *Org Geochem.* 24, 661-669.
Knicker, H. & Hatcher, P.G. (1997). *Naturwissenschaften* 84, 231-234.
Kovac, N.; Faganeli, J.; Bajt, O.; et al. (1995). In: *Organic Geochemistry: Developments and Application to Energy, Climate, Environment and Human History*; Grimalt, J.O.; Dorronsoro, C.: (Eds.); A.I.G.O.A.: Donostia-San Sebastian, Spain, pp 1153-1155.
Kovac, N.; Faganeli, J., Sket, B.; et al. (1998). *Org Geochem.* 29, 1623-1634.
Kovac, N.; Bajt, O.; Faganeli, J.; et al. (2002). *Mar Chem.* 78, 205-215.
Kovac, N.; Faganeli, J.; Bajt, O.; et al. (2004). *Org Geochem.* 35, 1095-1104.
Kovač, N.; Mozetič, P.; Trichet, J.; et al. (2005). *Mar Biol.* 147, 261-271.
Kovač, N.; Faganeli, J.; Bajt, O.; et al. (2006). *Acta Chim Slov.* 53, 81-87.
Kršinić, F. (1995). *J Plankton Res.* 17, 935-953.
Lancelot, C. (1995). *Sci Total Environ.* 165, 83-102.
Leppard, G.G.; Heissenberger, A. & Herndl, G.J. (1996). *Mar Ecol Prog Ser.* 135, 289-298.
Logan, B.E.; Passow, U.; Alldredge, A.L.; et al. (1995). *Deep Sea Res II* 42, 203-214.
Maestrini, S.Y.; Breret, M.; Bechim, C.; et al. (1997). *Estuaries* 20, 416-429.
Malej, A. & Harris, R.P. (1993). *Mar Ecol Prog Ser.* 96, 33-42.
Malej, A.; Malačič, V.; Mozetič, P.; et al. (2001). Marine mucous and possible measures for the mitigation of consequences. Report National Institute of Biology, Piran, pp 1-82 (in Slovenian).
Manganelli, A. & Funari, E. (2003). *Ann Ist Sup Sanita* 39, 77-95.

Marchetti, R.; Iacomini, M.; Torri, G.; et al. (1989). *Acqua Aria* 8, 883-887.
Mari, X. & Burd, A. (1998). *Mar Ecol Prog Ser.* 163, 63-76.
Mari, X.; Rochelle-Newall, E.; Torréton, J.P.; et al. (2007). *Limnol Oceanogr.* 52, 808-819.
Mayer, L.M. (1999). *Geochim Cosmochim Acta* 63, 207-215.
McCarthy, M.D., Hedges, J.I. & Benner, R. (1993). *Chem Geol.* 107, 503-507.
McCarthy, M.; Hedges, J.I. & Benner, R. (1996). *Mar Chem.* 55, 281-297.
Mecozzi, M.; Acquistucci, R.; Di Noto, V.; et al. (2001). *Chemosphere* 44, 709-720.
Mecozzi, M.; Amici, M. & Cordisco, C.A. (2004). *Chem Ecol.* 20, 41-54.
Mecozzi, M.; Pietrantonio, E.; Di Noto, V.; et al. (2005). *Mar Chem.* 95, 255-269.
Meyers-Schulte, K.L. & Hedges, J.I. (1986). *Nature* 262, 61-63.
Milani, L. & Fonda Umani, S. (1992). In *Marine Coastal Eutrophycation*; Vollenweider, R.A.; Marchetti, A.; Viviani, R.: (Eds.); Sci Total Environ (suppl.); Elsevier Science, Amsterdam, NL, pp 569-578.
Mingazzini, M. & Thake, B. (1995). *Sci Total Environ.* 165, 9-14.
Mohar, B.; Kofol, R.; Kovač, N.; et al. (2007). In *ASLO 2007 Aquatic Sciences Meeting: Water rocks!: Book of abstracts*. Santa Fe: ASLO www.aslo.org/santafe2007, pp 123.
Müller-Niklas, G.; Schuster, S.; Kaltenböck, E.; et al. (1994). *Limnol Oceanogr.* 39, 58-89.
Myklestad, S.M. (1995). *Sci Total Environ.* 165, 155-164.
Najdek, M. (1996). *Mar Ecol Prog Ser.* 139, 219-226.
Najdek, M.; Degobbis, D.; Mioković, D.; et al. (2002). *J Plankton Res.* 24, 429-441.
Nguyen, R. T. & Harvey, H.R. (2003). *Org Geochem.* 34, 1391-1403.
Olianas, A.; Fadda, M.B.; Boffi, A.; et al. (1996). *Mar Environ Res.* 41, 1-14.
Penna, A.; Berluti, S.; Penna, N.; et al. (1999). *J Plankton Res.* 21, 1681-1690.
Penna, N.; Berluti, S.; Penna, A.; et al. (2000). *Water Sci Technol.* 42, 299-304.
Penna, N.; Capellacci, S.; Ricci, F.; et al. (2003). *Anal Bioanal Chem.* 376, 436-439.
Penna, N.; Capellacci, S. & Ricci, F. (2004). *Mar Pollut Bull.* 48, 321-326.
Pettine, M.; Puddu, A.; Totti, C.; et al. (1993). *Suppl. Notiziario. S.I.B.M.* 1, 39-42.
Pettine, M.; Pagnotta, R. & Liberatori, A. (1995). *Ann Chim.* 85, 431-441.
Pistocchi, R.; Cangini, M.; Totti, C.; et al. (2005). *Sci Total Environ.* 353, 307-316.
Pokrovsky, O.S.; Gelabert, A.; Viers, J.; et al. (2002). *Geochim Cosmochim Acta* 66 (Suppl.), A609.
Pompei, M.; Mazziotti, C.; Guerrini, F.; et al. (2003). *Harmful Algae* 2, 301-316.
Posedel, N. & Faganeli, J. (1991). *Mar Ecol Prog Ser.* 77, 135-145.
Poutanen, E.L. (1985). *Mar Chem.* 17, 115-126.
Precali, R.; Giani, M.; Marini, M.; et al. (2005). *Sci Total Environ.* 353, 10-23.
Radić, T.; Kraus, R.; Fuks, D.; et al. (2005). *Sci Total Environ.* 353, 151-161.
Ramaiah, N.; Yoshikawa, T. & Furuya, K. (2001). *Mar Ecol Prog Ser.* 212, 79-88.
Revelante, N. & Gilmartin, M. (1991). *J Exp Mar Biol Ecol.* 146, 217-233.
Rinaldi, A.; Vollenweider, R.A.; Montanari, G.; et al. (1995). *Sci Total Environ.* 165, 165-183.
Russo, A.; Maccaferri, S.; Djakovac, T.; et al. (2005). *Sci Total Environ.* 353, 24-38.
Schnaitman, A.C. & Klena, J.D. (1993). *Microbiol Mol Biol Rev.* 57, 655-682.
Sellner, K.G. & Fonda Umani, S. (1999). In *Ecosystems at the Land-Sea Margin: Drainage Basin to Coastal Sea*; Malone, T.C.; Malej, A.; Harding, L.W.Jr.; Smodlaka, N.; Turner, R.E.: (Eds); Amer. Geophys. Union Publ., Washington, DC, pp 173-206.
Sihombing, R.; Greenwood, P.F.; Wilson, M.A.; et al. (1996). *Org Geochem.* 24, 859-873.

Staats, N.; Stal, L.J. & Mur, L.R. (2000). *J Exp Mar Biol Ecol.* 249, 1-12.
Stachowitsch, M.; Fanuko, N. & Richter, M. (1990). *P.S.Z.N.I: Mar Ecol.* 11, 327-350.
Stumm, W. & Morgan, J.J. (1996). *Aquatic Chemistry, Chemical Equilibria and Rates in Natural Waters*; John Wiley & Sons, Inc: New York, US, 3rd ed., 1022pp.
Supić, N. & Orlić, M. (2000). *J Marine Syst.* 20, 205-229.
Svetličić, V.; Žutić, V. & Hozić Zimmermann, A. (2005). *Annals of the New York Academy of Sciences* 1048(1), 524-527.
Thornton, D.C.O.; Santillo, D. & Thake, B. (1999). *Mar Pollut Bull.* 38, 891-898.
Tomasino, M.G. (1996). *Ecol Modell.* 84, 189-198.
Totti, C.; Cangini, M.; Ferrari, C.; et al. (2005). *Sci Total Environ.* 353, 204-217.
Verdugo, P. (1994). *Adv Polym Sci.* 110, 145-156.
Verdugo, P.; Alldredge, A.L.; Azam, F.; et al. (2004). *Mar Chem.* 92, 67-85.
Viviani, R.; Boni, L.; Cattani, O.; et al. (1995). *Sci Total Environ.* 165, 193-201.
Vollenweider, R.A. & Rinaldi, A. (Eds.) (1995). Marine Mucillages, with special reference to mucilage events in the northern Adriatic Sea, the Tyrrhenian Sea and North Sea. *Sci Total Environ.* 165, 1-230.
Volterra, L. (1995). *Sci. Total Environ.* 165, 225-228.
Weiss, M.S.; Abele, U.; Weckesser, J.; et al. (1991). *Science* 254, 1627-1630.
Wilson, M.A.; Barron, P.F. & Gillam, A.H. (1981). *Geochim Cosmochim Acta* 45, 1743-1750.
Zang, X. & Hatcher, P.G. (2002). *Org Geochem.* 33, 201-211.
Zegouagh, Y.; Derenne, S.; Largeau, C.; et al. (1999). *Org Geochem.* 30, 101-117.
Ziervogel, K.; Karlsson, E. & Arnosti, C. (2007). *Mar Chem.* 104, 241-252.
Žutić, V. & Svetličić, V. (2000). In *The Handbook of Environmental Chemistry* Part D; Wangersky, P. (Ed.); Marine chemistry: Springer-Verlag: Berlin Heidelberg, Vol. 5, pp. 150-165.
Žutić, V.; Svetličić, V.; Ivošević, N.; et al. (2004). *Period Biol.* 106, 67-74.

Chapter 6

GEOCHEMICAL SIGNALS AND PALEOCLIMATE CHANGES IN A 16,000 ^{14}C YEAR SEDIMENTARY RECORD FROM LAKE GUCHENG, EASTERN CHINA

R.L. Wang[1,*], *S.C. Zhang*[2,*], *S.C. Brassell*[3], *D. Tomasi*[1], *S.C. Scarpitta*[4], *G. Zhang*[5], *G.Y. Sheng*[5], *S.M. Wang*[6] *and J.M. Fu*[5]

[1] Medical Department, Brookhaven National Laboratory, Bldg. 555A, Upton, New York 11973, USA
[2] The Key Laboratory of Petroleum Geochemistry, PetroChina, P.O. Box 910, Beijing 100083, PRC
[3] Biogeochemistry Lab, Department of Geology, Indiana University, IN 47405, USA
[4] SCDHS, Hauppauge New York 11788, USA
[5] The State Key laboratory of Organic Geochemistry, Institute of Geochemistry, Chinese Academy of Sciences, Guangzhou 510640, PRC
[6] Nanjing Institute of geography and Limnology, Chinese Academy of Sciences, Nanjing 210008, PRC

ABSTRACT

A long sedimentary core (20 m) taken from Lake Gucheng, Nanjing, Jiangsu Province of eastern China was studied focusing on geochemical changes versus paleoclimatic and hydrological history. ^{14}C dating results of fossil organic carbon from the GS-1 core indicate the oldest sample was deposited around 16 kyr BP. The location of this lake at the intersection of monsoon climate regions in continental China provides a valuable opportunity to study the Late Quaternary climate history. Organic and inorganic carbon contents, stable carbon and oxygen isotopic compositions (δ^{13}C and δ^{18}O values), trace metal elements, coupled with indicative mineral contents, provide detailed comprehensive geochemical signals suggesting marked climate and hydrological changes in this region since the end of last glacial maximum.

---

[*] corresponding authors: e-mail: rlwang@bnl.gov and sczhang@petrochina.com.cn

Sediment samples from the lower part (19.7 – 12.5 m; 15.9 – 10 kyr BP) of the cores are characterized by higher contents of quartz and albite, the relatively weatherproof minerals, and lower in carbonate and clays, the apparently deeper water deposits. Organic and inorganic carbon, trace metal elements (Cr, Co, Zn, Pb, Ni etc.) and Na/Fe ratios are lower than those of the upper part of the cores (12.5 – 0 m; 10 kyr BP - recent), suggesting a deeper lake environment developed following the end of the Last Glacial maximum and the enhanced monsoon precipitation and warming up in the region starting at the beginning of Holocene. Stable isotope data, however, shows more fluctuations rather than the two clearly cut stages between the Late Pleistocene and the Holocene as shown by the mineral, elemental and lithological data. These isotope data show that several important paleoclimatic events including the ending of the Last Glacial Maximum (16 - 15 kyrBP), the severe cold/dry event Younger Dryas (10 - 11 kyrBP), the optimum climate during the mid-Holocene, and a substantial cooling and drought episode during the Iron Age Neoglaciation in later Holocene (3500 - 2000 yrBP), may all be observed with geochemical signals in this inland lake despite the overlapping complications of the other no-climatically induced geochemistry effects such as sediment diagenesis and hydrological environment changes.

Key Words: Pleistocene, Holocene, climate change, stable isotope, lacustrine geochemistry, Younger Dryas, Holocene Climate Optimum, Neoglaciation.

1. INTRODUCTION

Holocene paleoclimate history of eastern China is of massive interest since its course has more or less shaped the long agriculture dominated Chinese civilization started thousands of years ago. Flood and draught of the two main rivers, Yellow and Yangtze and five major lakes has directly influenced this region broadly [1]. Lake Gucheng, Nanjing (Figure 1) is located in the heart of this region that is believed to be sensitive to all crossing climate consequences including summer and winter monsoons [2]. Therefore, this lake has been chosen as a local model site for late Quaternary studies [2-3]. Stable isotope records ($\delta^{13}C$ and $\delta^{18}O$) of lacustrine sediments can reflect the kinetic history of carbon and oxygen in the lake, including history of photosynthesis, precipitation, surface/subsurface input and output and are valuable evidence for the paleoclimate change in the continental environments [4-8]. The $\delta^{18}O$ values of precipitation are controlled by several climatic factors such as water temperature, isotopic fractionation between rainout and airmass, which controls the material balance of the lake in general. The isotopic composition of lake water is generally controlled by isotopic equilibrium between the total input and output of waters. The $\delta^{18}O$ values of sedimentary minerals in lacustrine sediments thus can reflect the $\delta^{18}O$ value of precipitation and therefore can be used as an indicator of regional or even hemispheric or global paleoclimate [e.g., 4-9].

The studied lake (Figureure 1) is located in one of the key areas sensitive to climate change associated with summer and winter monsoon precipitation history in the northern hemisphere [10-12]. Paleoclimate records with age-controlled sediment cores in this region are inadequate. Studies on the lacutsrine sedimentary records could be helpful for understanding the possible relationship between lacustrine geochemistry and regional climate history [4-9], mechanisms of climate change in this region. Here we present our recent effort to use a 20 m sedimentary core taken from Lake Gucheng of Jiangsu Province of eastern China (Figure 1).

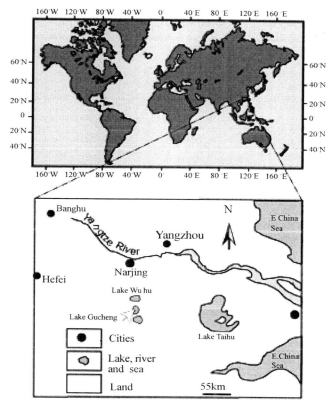

Figure 1: Maps showing the location of Lake Gucheng in Jiangsu Province, China in the background of the world map marked with longitude and latitudes (31 14'36"-31 19'28"N 118 51'34"-118 57'56"E).

2. GEOLOGICAL SETTING AND REGIONAL CLIMATE

Lake Gucheng is located in the Gaochun County of Jiangsu Province (31°14'36" - 31°19'28"N; 118°51'34" - 118°57'56"E). The lake has an area of about 31 km^2, average water depth of 1.56 m and maximum water depth of 6.79 m. This region is located within the northern-middle subtropical monsoon climate region with warm/mild climate (annual average temperature is 15.9°C) and abundant summer precipitation (annual precipitation of 1125.7 mm, Table 1). Core GS-1 was drilled near the west bank of the lake (Figure 1) [2].

Table 1 shows the general features of Lake Gucheng and the vicinity region. Due to the low evaporation/precipitation ratio (E/P = 0.12, Table 1), the lake water is currently fresh (Table 1) [2].

The average elevation of the lake is about 6 m above sea level. During the Pleistocene climatic cooling (Last Glacial Maximum, LGM), sea level was > 100 m lower than the present, coastal line of eastern China Sea located at the edge of the continental shelf more than 1000 km away from today's coastal line [13]. Lake formation of Gucheng was initiated by the deep incision of the Yangtze River system during the LGM [13].

Table 1: General geographical and hydrological data of Lake Gucheng [9]

Average Depth (m)	1.56
Maximum Depth (m)	6.79
Area (km^2)	31
pH	8.2
Annual average temperature (°C)	15.9
Annual precipitation	1157 mm
Annual evaporation (mm)	144.3
Salinity (%)	< 0.05
E/P ratio	0.12

3. METHODOLOGIES

3.1 Measurement of Elemental Carbon and Sulfur Concentrations

The core, 20 m long, was separated into sections and 50 samples were freeze-dried to remove the water at the Institute of Geochemistry of CAS prior to shipment to the US. Elemental analyses of total carbon (TC), total organic carbon (TOC) and total sulfur (TS) contents were performed on homogenized dry samples using a LECO C/S Analyzer [9]. Total carbon was determined by combustion samples at 1600°C and subsequently converted by the measured CO_2 using infrared detection. TOC contents were obtained in the same way using the residual samples after treatment with 1M HCl. Total sulfur (TS) and total residual sulfur (TRS, including total organic sulfur and the residual sulfides such as pyrite sulfur etc.) were measured using the same analyzer by combustion and measurement of SO_2. Total inorganic carbon (TIC) and total dissolvable sulfur (TDS) are the difference between TC and TOC, TS and TRS, respectively.

3.2 Isotopic Measurement of Carbonate Sediments

Stable isotope ratios of carbon and oxygen on bulk carbonate were measured on a Finnigan Mat MS-252 mass spectrometer using standard analytical procedures. CO_2 from carbonate samples was generated by reaction with 100% phosphoric acid using the inverted y-tube technique at 25°C. The isotopic composition of samples is defined and reported in the conventional δ-scale in parts per thousand:

$$\delta_{sample} (‰) = [(R_{sample} - R_{standard})/(R_{standard})] \times 1000$$

where R is the abundance ratio of $^{13}C/^{12}C$ or $^{18}O/^{17}O$ in the sediment samples or in the standard. Isotope ratios are reported relative to the vPDB (Pee Dee Belemnite) standard for carbon and oxygen isotope measurements on generated bulk CO_2 [6-9].

3.3 Trace and Major Metal Measurement by Inductively Coupled Plasma Mass Spectrometry

Dried sediment samples were weighed and put in glass and concentrate reagent grade nitric acid was added to the open vials. Vibration of the reaction was carried out on a Maxi Mix II mixer and then the solution was placed on a hotplate at a temperature ~ 95°C for 2 x 8 hr and continuously mixed by a magnet mixer. When the digestion was completed, the solution was transferred quantitatively into a volumetric flask and brought up to level by adding 5% nitric acid in type I reagent water. The volumetric flask was kept for further analysis. The top clear level of the solution was quantitatively diluted for ICP-MS analysis. An HP-4500 ICP-MS instrument was used in the analysis.

4. RESULTS

4.1 Radiological Dating Results of Samples

Totally 11 points were obtained for the 20 m long core from Lake Gucheng as shown in Table 2 together with their burial depth. Seven of these points are AMS ^{14}C dating and others are conventional ^{14}C dating results (Figure 3) all by the ^{14}C Lab at Institute of Geochemistry Chinese Academy of Sciences. Extracted kerogen carbon or/and mollusks were used the ^{14}C measurement. All the points were used in the mathematical regression in Figure 3. These data show that the average sedimentation rate of the lake is about 1.1 mm/year (Figure 3), much higher than the saline lake sedimentation rate of ~ 0.24 mm/year [14]. For the regression fitting in Figure 3, constant sedimentation rate between adjacent radiocarbon dates is assumed.

Table 2 ^{14}C radiometric date of core samples from core GS-1 of Lake Gucheng, E. China

Sample	Depth (m)	^{14}C age (aBP)
GS36	3.6	3,060 ± 420
GS73	5.42	2,500 ± 350
GS126	8.10	5,680 ± 710
GS144	9.00	5,000 ± 520
GS220	12.37	8,010 ± 680
GS276	15.35	12,300 ± 560
GS282	15.80	14050 ± 600

Numerical regression of age-depth curve based on these ^{14}C dates is illustrated in Figure 3 using the mean depth (m) vs. mean age of samples using Cricket Graph III by Computer Associate Inc.

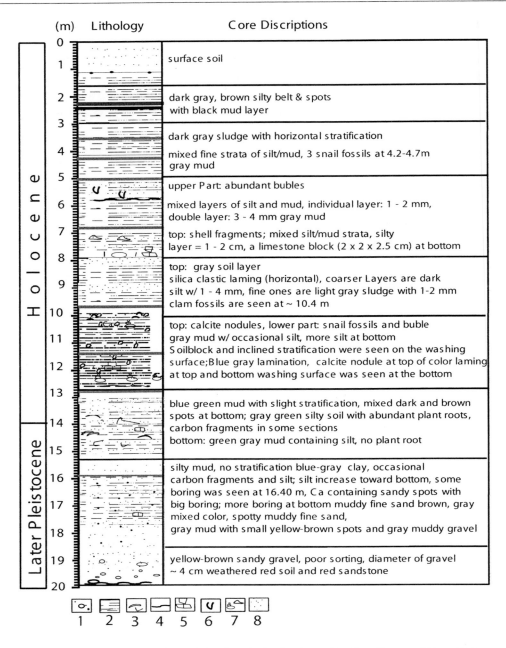

Figure 2: Lithological column of core GS-1 from Lake Gucheng, Nanjing, Eastern China (Pleistocene and Holocene boundary is suggested based on depth-age relationship as interpolated and inFigure 3; 1- sandy gravel; 2-laminated slit; 3- plant root/carbon fragments; 4-wash surface; 5-limestone blocks; 6- sediment with bubble; 7-shell or shell fragments 8-soil; Section 1-4 after Wang et al. 1996)

4.2 Carbon and Sulfur Elemental Concentrations

Organic carbon, inorganic carbon and sulfur concentrations for the GS-1 core samples are plotted in Figure 4. Sulfur content of the cores are constantly low, total sulfur remains at ~

0.25% of the total dry sediments. The total sulfur content of the sediments remains quite stable, indicating that the salinity change of the lake water was insignificant (Figure 2), even though sulfur cannot be simply used as paleo-salinity proxy due to other geochemical affects. Organic and inorganic carbons show notable changes with a marked positive shift of carbon contents in the cores at around 12 m (10 kyrBP). Organic carbon is low, 0.25% in the section I, 20 - 13.7 m, but increases to 0.35% at the top of this section and then continues to rise from 0.35% to 0.6% and remains quite stable at this level throughout the section with only minor fluctuations (Figure 4). Meanwhile, inorganic carbon shows a substantial increase around 12 m (10 kyrBP) from about 0.25% to 0.5% and continues to increase to its peak value of about 1.0% at about 4 m.

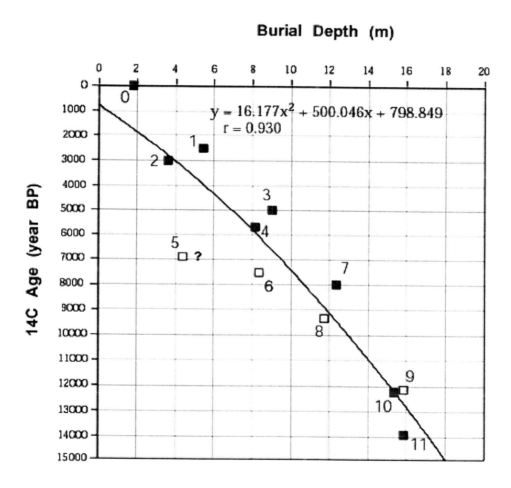

Figure 3: Cross plot of 14C radiometric dates and burial depth of core GS in Lake Gucheng, Eastern China. (The AMS radiocarbon dates (solid except #0) were measured on total organic carbons at Institute of Geochemistry, CAS; dates # 5, 6, 8 and 9 are previously published data after Wang et al., 1996 [2]; all used in the linear regression shown by the equation in th plot; point 0 is the starting time (1950) of landfill, 50 yrBP was assumed for this point; mean burial depth (m) and mean ^{14}C date of each sample were used in the plot. Point 5 might be due to the contamination from some re-deposited materials as also seen in the Figure 2).

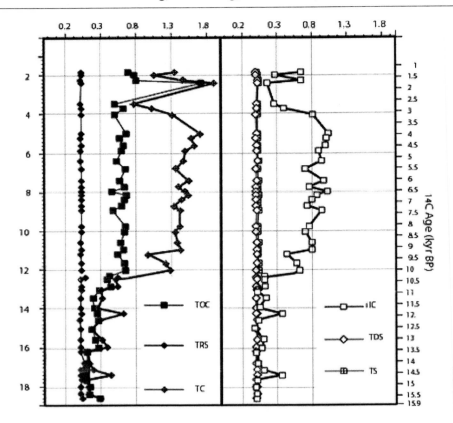

Figure 4: Distribution of total organic and inorganic carbon and sulfur concentrations of Core GS-1 from Lake Gucheng, Nanjing, East China (concentrations are reported in % in dry weight sediments; TOC-total organic carbon; TRS-total residual sulfur; TC-total carbon; TIC-total inorganic carbon; TDS-total dissolvable sulfur; TS-total sulfur).

4.3 Mineral Compositions

Mineral compositions are plotted against burial depth in Figure 5. The bottom section (20 -13 m, Later Pleistocene) is characterized by higher contents of quartz and albite and low in clay minerals and carbonate. From about 12 m and up sediments are dominated by clay and carbonate. Higher concentration of montomorillonite, illite, kaolin and chlorite probably suggests the development of lower energy sedimentation began at around 10 kyrBP. Quartz and albite dominate the bottom section (20 - 12 m) of the core. These two minerals drop rather quickly in the sediments above 12 m and reach the minimum levels at around 9 m. Quartz remains in low level, until a marked increase at ~2.5 m. Albite shows moderately high level at about 6 - 7 m. Carbonate contents (calcite and dolomite) are low in the lower section (Pleistocene) and they reach the maximum level of the whole interval at ~7.5 and 8.25 m, perhaps an indication of formation of brackish environment in the middle of Holocene.

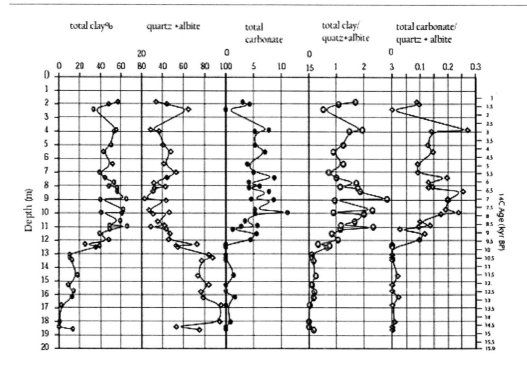

Figure 5: Distribution of minerals in the sediments from core GS-1 in Lake Gucheng, East China (relative mineral compositions are normalized into percentages; measurements were carried at Inst of Geochemistry, Chinese Academy of Sciences; total clay = montmorillionite + illite + chlorite + kaolin; total carbonate = calcite + dolomite).

4.4 Metal Concentrations

Heavy metals and other trace elements and their ratios are shown in Figure 6. Rapid change of trace and major elements in the core can be seen at the depth around 13.7 m. Section 20-13.7 m is characterized by low level of metals. Abundance of these elements reflects the dramatic change during the past 16,000 years, due to either natural or even anthropogenic influences in their sedimentary accumulation rates in the lake. Further discussions will be opend in the Discussions section of the text.

4.5 Carbon and Oxygen Stable Isotope Values ($\delta^{13}C$ and $\delta^{18}O$)

Figure 7 illustrates the distribution of stable isotope values of carbon and oxygen in the sediments. Distributions of $\delta^{13}C$ and $\delta^{18}O$ values along the 20 m core will be presented separately in the following two subsections.

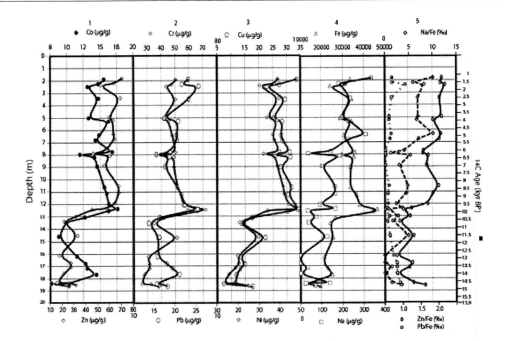

Figure 6: Distribution of trace metal contents (column #1-4) and ratios of Na, Zn and Pb to Fe (column #5, of core GS-1 from Lake Gucheng, East China. All the concentrations are in micrograms per gram of 1 dry sediment.

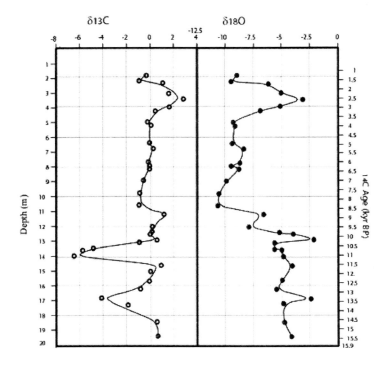

Figure 7: Distribution of $\delta^{13}C$ and $\delta^{18}O$ values of bulk sediments from Core GS-1 of Lake Gucheng, East China (isotope values are in ‰ vPDB.

4.5.1 Carbon Isotope Values ($\delta^{13}C$)

$\delta^{13}C$ values of the whole sequence span a rather large range from -7 to +3‰. Core section 20 – 14 m is characterized by high values of $\delta^{13}C$ (-1 to +1‰), except for one sample at ~ 17 m ($\delta^{13}C$ = -4‰). A negative shift of $\delta^{13}C$ values occurs at the depth around 14 –13 m, from the regular range of -1‰ to as low as -8‰. From about 13 m to 11 m, the $\delta^{13}C$ values shift positively back to the range of -1 ~ +1‰. $\delta^{13}C$ values of this section remains around a balanced value (~0 ‰) except for the one low point at 10.5 m ($\delta^{13}C$ = -1‰) and one high point at 11 m ($\delta^{13}C$ = +1‰, Figure 7). Core section 4 - 2.5 m is characterized by a higher value of $\delta^{13}C$ (+1 to +3‰). The top of the sampled core (2.5 m and above) is characterized by a quick dropping of $\delta^{13}C$ values by 2‰, from +3 to -1‰ (Figure 7).

4.5.2 Oxygen Isotope Values ($\delta^{18}O$)

Oxygen isotope values ($\delta^{18}O$) also have a wide range from -2 to -11‰ (Figure 7-b). The bottom section is characterized by higher $\delta^{18}O$ values (–3‰ to -6‰), with two exceptions of $\delta^{18}O$ values above –2.5‰ at depth of ~17 m and ~13 m, respectively. A sharp drop occurs from 13 to 11 m taking the $\delta^{18}O$ values from about –2‰ to the lightest point of the whole sequence of –11 ‰, except for a minor fluctuation at 11.5 m elevating the $\delta^{18}O$ value up to -7‰. The $\delta^{18}O$ values of the core section 11 - 5 m are the lowest among the whole sampled cores, slightly fluctuating from –11 to –9‰, averaging at around –9.5‰. A rather sharp rise of $\delta^{18}O$ values is observed in section of 4 - 2.5 m, bringing the average $\delta^{18}O$ values up to around -5‰. However in the sections above 2.5 m, both carbon (as shown in text of 4.5.1) and oxygen isotope values drop from their peak values to a much lower level (e.g., $\delta^{18}O$ values shift from -2‰ to –9.5‰).

5. DISCUSSIONS

Even though lacustrine sediments may have a complicated origin and may have undergone varieties of geochemical changes after deposition, the overall trend of the geochemical signals we observed in the Lake Gucheng appears to be climate/hydrologically controlled. Thus, these geochemical signals could be used as paleoclimate proxies with certain caution.

Figure 8 combines the mineral, elemental and isotopic data as well as the previously published data on pollen counting results from the core GS-1 [2]. Interpretations in the paleoclimatic signals are summarized in the right side of the figure in order to simplify discussions.

The distribution of minerals reflects the strength of weathering of base rocks from the drainage area and also the energy level as well as the hydrological status of the water system of the lake. Quartz is one of the most weather-proof (chemical or mechanical) minerals and it also in general reflects a relatively higher energy level of the lake hydrological status in comparison to mud/clays and carbonates or silty carbonate [15]. During the early stage of the lake, i.e. later Pleistocene, sedimentation was deposited mainly from the more dynamic water (fluvial) from de-glaciation. Carbonate and clay minerals are both extremely low in this section of the sediment core but they increase rapidly above 14 m with predominance of low energy lacustrine phase formation around 11 kyrBP (Figure 2 and Figure 5), Assessment of natural sources of atmospheric trace metals on a global scale for present climatic conditions (interglacial conditions) indicate that atmospheric metals originate from wind-born dust (soil) and rock particles, seasalt spray, volcanoes, wild forest fires, and continental and marine biogenic particulates and volatiles [16-19]. Like the large changes in the concentrations of heavy metals in later Pleistocene and Holocene ice cores [16-19], the pronounced variation of metals in Lake Gucheng sediments may also be due to large changes of the regional or even hemispheric environment. The dramatic raise of metals seen at the transition of Pleistocene to Holocene in the sediment core of lake Gucheng can be largely related to the glaciation and deglaciation cycles, which in turn affect the vegetation coverage of the dust source area (northern China Losses Plateau, northwestern deserts) [10-12].

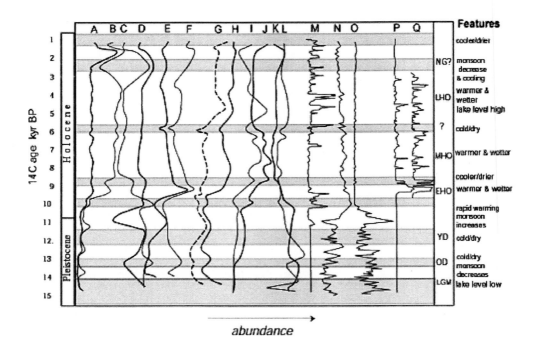

Figure 8: Summary of different paleoenvironmental proxies and the interpreted paleoclimatic scenarios for the past 16kr ^{14}C years. (Pollen/spore diagrams after [2]; Stages and chronozones after [8], alphabetic designations are: A-D: total organic carbon, total inorganic carbon, $\delta^{18}O$, $\delta^{13}C$, E-H: Cr, Pb, Na/Fe and Zn/Fe ratios; I-L: chlorite, kaolinite, quartz, albitem; and M-Q: pollen and spore, wood and grass pollen, *Artemisia*, total diatom, and Coscinodiscus *sp.* abundance, respectively. Climatic events: NG: Neoglaciation; E/M/LHO: early, middle and late Holocene Optima resectively; YD: Younger Dryas, OD: Oldest Dryas; LGM: last glacial maximum; grey zones indicate the colder climate).

Carbon and oxygen isotopic signals extracted from core GS-1 suggest that significant changes in climate as well as the climate-related paleoenvironmental characteristics occurred at around 10000 years in this region. Even though LGM ended at around 15 kyrBP, the 5 kyr after LGM was still rather cold and dry in this region until the beginning of Holocene Epoch ~10 kyr BP.

5.1 20-12.5m (15.9-10 ^{14}C kyrBP, Later Pleistocene)

This stage is equivalent to the later glacial stage, including Oldest Dryas, Bølling, Allerød and Younger Dryas [4,8]. Cores are low in TOC and TC (Figure 4), trace metals (Figure 6) and total pollen counts are also low (Figure 8, [2]. Carbon and oxygen isotope values are high, except the middle section (Allerød). Minerals are dominated by quartz and albite, low in kaolin and chlorite and other clay minerals (Figure 5). Heavy metals are particularly low in this section of sediments (Figure 6).

5.1.1 20-17.4m (15.9-13.7 ^{14}C kyrBP, Oldest Dryas, Bølling)

TOC and TC are low; quartz and albite are high, while chlorite and other clay minerals are minimum. Sediments are enriched in heavier isotopes that both δ^{13}C and δ^{18}O values of bulk sediments are >0 ‰ (Figure 7). ICP-MS data shows that both heavy metals and other trace metals are low in these sediments (Figure 6). According to an earlier study [2], wood pollen is low in the pollen counts, dominated by *Pinus, Quercus, Betula, and Ulmus*. Others like *Cyclobolanopsis, Castanea, Ilex, Linguidambar, Carya* were not identified in the samples. Pollen abundance is low (5-200 /cm^3). Diatom pollen is extremely low. These lines of evidence all point to a cold and arid environment with low lake level, low bioproduction in the lake and its drainage area. This region was perhaps still a 'pre-lake', or fluvial environment. According to the COHMAP Members [19], at round 15 kyrBP, summer solar radiation significantly increased. Summer wind activity increased in this region and caused monsoon precipitation increase significantly. The beginning of Lake Gucheng is the result of enhanced monsoon precipitation following the global deglaciation.

5.1.2 17.4 - 16.1m (13.7 - 12.7 ^{14}C kyrBP)

An obviously warmer and wetter millennium with enhanced temperature and water input to the lake. Carbon isotope values drop to –4% but δ^{18}O values increase to +2%. Metals increase rapidly comparing to the last stage.

Palynological data [2] shows that wood pollens increase in this section of the core, dominated by *Cyclobolanopsis*, Quercus, Junglans and occasionally Diatom are still low, but rather diversified. However, TOC and TC remain low and clay minerals increase relative to quartz and Albite.

5.1.3 16.1 - 14.1m (12.7 - 11.1 ^{14}C kyrBP)

TOC and TC are low; especially the middle section of this subunit (<0.3%), indicating biological production was low. Both carbon and oxygen isotope values have a marked positive shift (Figure 7 and 8). Trace elements and major metals decrease and pollen and spore counting remain low [2]. All these lines of evidence seem to point to a colder and severely drier environment, equivalent to the Younger Dryas event in Europe and North Atlantic region [4, 6-8]. A decrease of about 2‰ at the Allerød/Younger Drays transition is interpreted as a drop in mean annual air temperature of approximately 5°C. An abrupt temperature increase of a similar magnitude is inferred at the Younger Dryas/Preboreal boundary [7-9]. Data from southern China Sea showed that domination of *Picea, Abies* and *Tsuga* pollen around 11000 years BP in this region, indicating a cool but wet climate for southern China throughout the Younger Dryas chronozone [21]. The pollen counting (mountain conifer species) increase to a maximum level at about 10600 ^{14}C yrBP (12500 cal yrBP), then sharply drop towards the Younger Dryas/Holocene boundary. After all, such a sharp change of stable isotope and element compositions at P/H boundary in Lake Gucheng is rather similar to what was previously reported in the western Tibetan salt lakes [7, 9].

5.1.4 14.1-13.1 m (11.1-10.25 ^{14}C kyrBP)

TOC increases obviously, indicating probably an enhanced production/preservation during this period of time. Palynological data [2] shows that wood pollen rapidly increases and grass pollen also increases obviously. δ^{13}C data is rather negative and drops to the lowest level of the whole core (δ^{13}C~-6.5‰). Perhaps this negative excursion of carbon isotope values could be the marking point for the beginning of Holocene Epoch. δ^{18}O values of carbonate also dropped negatively during this period although it is not as remarkable as that of carbon. δ^{18}O shift probably indicates that the large quantities of fresh water (meteoritic precipitation and melting water from upstream 'ice sheet') happened rapidly in the region. Increased monsoon precipitation and global warming after the last glacial period increased the global weathering area, releasing large quantities of 'fresh' CO_2 input to the atmospheric pool [22]. The ultimate result is the sharp decrease of δ^{13}C values at the beginning of the Holocene (Δ^{13}C >7‰).

5.1.5 13.0-12.5 m (10.25-9.7 ^{14}C kyrBP)

Rapid positive shift of carbon and oxygen isotope values marks the interval (Δ^{13}C >7‰ and Δ^{18}C >3‰, Figure 7 and 8). Wood and grass pollen counts are minimum and the total pollen and spore abundance drops markedly right after the beginning of this interval. These lines of evidence suggest that the climate was rather cold and drier. Concentration of the water due to lack of monsoon precipitations caused the enrichment of heavier isotopes in the lake water and then the sediments. The cold and dry climate is also reflected in the low pollen and spore abundance and organic carbon values in the sediments. Decrease of monsoon precipitation could be responsible for the low level of trace elements in the sediments (Figure 8).

5. 2 12.5 - 6.85m (10 - 5.3 ^{14}C kyrBP, Early-Mid Holocene)

5.2.1 12.5 - 11.2 m (9.7-8.8 ^{14}C kyrBP)

This interval is characterized by the rapid increase of TOC and total carbon, trace elements and major metals in the sediments. δ^{13}C values remain stable, δ^{18}O values first shift negatively from −2.5‰ to −7.5 ‰ and then increases slightly to ~6.5‰. Pollen and spore abundance, including wood, grass diatom and other algal species increased markedly. Clay minerals increase largely relative to quartz and albite (Figure 5). Carbonate minerals increase markedly relative to other mineral materials (Figure 5). All these lines of mineral and elemental evidence indicate Lake Gucheng was probably gradually developing into a brackish environment. Fluvial environment ended rapidly with termination of Pleistocene. It is reasonable to conclude that these ~1000 years were obviously warmer and wetter than today's climate.

5.2.2 11.2-10.6 m (8.8-8.3 14C kyrBP)

Obviously lower TOC, pollen counts, slightly dropped trace and major elements (Figure 6) are seen in this section of cores. As well observed are the slightly elevated $\delta^{13}C$ and $\delta^{18}O$ level ($\Delta \sim$ 1‰, Figure 7). Relative abundance of clay minerals is also lower than that of the underlying subsection (Figure 5). All the lines of evidence indicate that these ~500 years were colder and drier period. This period of dry episode was broadly reported in the northern Hemisphere including [e.g., 6, 33].

5.2.3 10.6-8.3m (8.3-6.4 ^{14}C kyrBP)

After the last short (~ 500 years) climate recession, Holocene climate evolved into its first major stable period of optimum condition. $\delta^{13}C$ and $\delta^{18}O$ levels dropped rapidly from the last period of heavy ($^{13}\Delta$ and $^{18}\Delta \sim$ 2‰ and 4‰, respectively, Figure 7), then they began to increase slowly and remained rather stable until the end of this period. The rapid drop of stable isotope values from the peak of last interval indicates a rapid increase of fresh input of water (and [CO_2], indicting the rapid increase of monsoon precipitation. Enhanced climate conditions resulted in risen primary production of the lake as signaled by elevated TOC values and total pollen counts (Figure 8). They both remain high throughout this interval. Sediment itself also provides supporting evidence of higher lake level resulting in increase of fine-grained minerals (clays) and chemical precipitation (carbonate) to a peak level of the whole Holocene. Carbonate/quartz ratio maximizes at around 10 and 8.3 m (7.8 and 6.5 ^{14}C kyrBP, respectively).

Trace element concentrations decrease slightly in this interval, perhaps due to the full expansion of vegetation coverage in the drainage area. Na/Fe ratios (Figure 6), however, increase markedly since the beginning of Holocene, an indicator of increase of salinity. Selective biological consumption of elements, increases of biological consumption of iron relative to other elements (e.g. Na) might have also played a major role in the lake and the vicinity drainage area (Figure 6) [23, 24].

5.2.4 8.3-7.8m (6.4-6.0 ^{14}C kyrBP)

The only short climate recession seems to have occurred during the Early-Mid Holocene at around 6~6.4 kyrBP. This short cold/dry pulse occurred between the two extended optima of Holocene climate periods. Major changes are observed in the drop of trace and major elements (Figure 6) and slight drop of total organic carbon as well as a brief slight negative shift of $d^{18}O_{carb}$ (Figure 7).

5. 3 6.85 – 2.5 m (6.0-1.2 ^{14}C kyrBP, Early-Mid Holocene)

5.3.1 6.85 - 4.8m (6.0-3.6 ^{14}C kyrBP)

Similarly, this was a major stabilized warm/wet interval of Holocene. All the lines of evidence, isotopic, mineral, elemental and palynological, all suggest that these two and half millennia were under an optimum condition. Wood pollen dominates (60-90%), indicating that during Middle Holocene, vegetation of the Lake Gucheng region was dominated by forest that replaced the grassland of the later Pleistocene and early Holocene. These wood pollens are dominated by *Pinus massoniana* (which is common in the sub tropic region of central-north Asia), *Cyclobolanopsis, Quercus, Linguidambar* and *Castanea* [2]. In the section of 12.9m and upper Diatom is abundant, suggesting warmer and more productive environment developed, in good agreement with other geochemical proxies (Figure 8).

5.3.2 4.8-2.8 m (3.6-2.0 ^{14}C kyrBP: Iron Age Neoglaciation)

Total organic carbon and total carbon values drop from their peak level of Middle Holocene to the lowest level at ~3.3m (~2.5 ^{14}C kyrBP). Meanwhile, carbon and oxygen isotope values increase quickly from their low values of Middle Holocene to the highest level of Later Holocene at the same depth (3.3m, 2.5 ^{14}C kyrBP). Heavy metals and trace element concentrations also rise to a new high level although the ratios to iron remain unchanged. Sediment minerals remain unchanged during these two millennia. Total pollen abundance decreases, including wood pollens (*Cyclobolanopsis, Linguidambar*) and water plants (*Typha, Myriophyllum,* and *Potamoge tonaceae*) [2]. All these lines of evidence suggest that during these two millennia the climate was obviously colder and drier than the previous stage. Monsoon precipitations were obviously lower, especially at around 2.5 ^{14}C kyrBP. This period of marked colder and arid time coincides with the Iron Age Neoglaciation as was reported in Europe and North America previously [25,26]. These data suggest that this cold and arid period was a global event in later Holocene for the vast area of northern hemisphere. Recent study on Lake Ahung Co in central Tibet [27] shows there was a severe late Holocene dry period that occurred around 4700 yrBP due to a rapid decline of monsoon precipitation. Using the annual varved lake sediments, Dean at al. [28] found a severe dry period around 4000 yrBP and a dry pulse about 600 year ago. Studies on sediment grain size from Lake Daihai [28] also suggest the general decrease of precipitation in northern China after 3100 years.

5.3.3 2.8-2.5 m (2.0-1.2 ^{14}C kyrBP)

Isotope, elemental and mineral data from core CS-1 demonstrate that during these ~800 years (2.0-1.2 kyrBP), climate warmed up quickly from the previous cold/arid stage (INA). Increased monsoon precipitations and the elevated temperature brought up the lake production, as indicated by higher TOC and pollen abundance (Figure 8). Enhanced monsoon precipitation and warmer climate caused the fresh input of water and [CO_2] to the lake, resulting in the quick drop of both $\delta^{13}C$ and $\delta^{18}O$ values at around 1800 yrBP. The increased silty and median grain sediments denoted intensification of hydrological cycles in northern China during 1700-1000 yr BP [28].

5. 4 2.5 – 0m (after 1.2 ^{14}C kyrBP, Late Holocene)

From the previous warm and wet climate condition, climate cooled down and dried up once again. Evidence includes lower TOC and lower pollen counts (lower production), as well as heavier $\delta^{13}C$ and $\delta^{18}O$ values in the top section of the cores. According to Wang *et al* [2], total pollen abundance, especially water plants, dropped obviously. Significant increases of Pb, Cu, Zn and other trace metals in the sediments (Figure 6) can likely be attributed to an increased evaporation/precipitation rate owing to the draught condition. We think this cooling/arid episode is probably associated with the Little Ice Age (LIA, AD 1270-1850).

This cooling/arid climate event was thought to have enhanced a drought condition globally, especially in Africa and broad area of northern hemisphere including northern China [30-33] as well as the North America [33].

6. CONCLUSION

Organic and inorganic carbon contents, stable carbon and oxygen isotopic compositions ($\delta^{13}C$ and $d^{18}O$ values), trace metal elements, coupled with indicative mineral contents, provide comprehensive geochemical evidence suggesting marked climate and hydrological changes in this region since the end of last glacial maximum. Sediment samples from the lower part (19.7 – 12.5 m; 15.9 – 10 kyr BP) of the cores are characterized by higher contents of quartz and albite, the relatively weatherproof minerals, and lower in carbonate and clays, the apparently deeper water deposits. Organic and inorganic carbon, trace metal elements (Cr, Co, Zn, Pb, Ni etc.) and Na/Fe ratios are lower than those of the upper part of the cores (12.5 – 0 m; 10 kyr BP - recent), suggesting a deeper lake environment developed following the end of the Last Glacial maximum and the enhanced monsoon precipitation and warming up in the region starting at the beginning of Holocene. Stable isotope data, however, shows more fluctuations rather than the two clearly cut stages between the Late Pleistocene and the Holocene as shown by the mineral, elemental and lithological data. These isotope data show that several important paleoclimatic events including the ending of the Last Glacial Maximum (16 - 15 kyrBP), the severe cold/dry event Younger Dryas (10-11 kyrBP), the optimum climate during the mid-Holocene, and a substantial cooling and drought episode during the Iron Age Neoglaciation in later Holocene (3500 - 2000 yrBP), may all be observed with geochemical signals in this inland lake despite the overlapping complications of the other no-climatically induced geochemistry effects such as sediment diagenesis and hydrological environment change.

ACKNOWLEDGMENT

We thank Dr. Arndt Schimmelmann and another anonymous reviewer who provided many thoughtful suggestions and comments for the primary version of this manuscript. Lisa Muench and Phil Hayde (BNL) are also acknowledged for their assistance in part of the metal analysis. The study was partly supported by the Department of Energy (Office of Biological and Environmental Research under contract DE-AC02-98CH10886). Initial sampling and ^{14}C dating were supported (RLW and SCB) by Chinese Academy of Science (CAS) through The Key Lab of Organic Geochemistry of CAS. Notice: This manuscript has been authored by employees of Brookhaven Science Associates, LLC under Contract No. DE-AC02-98CH10886 with the U.S. Department of Energy. The publisher by accepting the manuscript for publication acknowledges that the United States Government retains a non-exclusive, paid-up, irrevocable, world-wide license to publish or reproduce the published form of this manuscript, or allow others to do so, for United States Government purposes.

REFERENCES

[1] Cheng-Bang A., Tang, L., Barton, L., et al. (2005). Climate change and cultural response around 4000 cal yr B.P. in the western part of Chinese Loess Plateau. *Quaternary Research* 63(3): 347-352.

[2] Wang, S.M., Xiangdong, Y. & Yan, M. (1996). Environmental Change of Gucheng lake of Jiangsu in the past 15 ka and its relation to palaeomonsoon, *Science in China* (D), 26: 137-141.

[3] Zhang G., Guoying, S. & Jiamo, F. (1999). The occurrence of hydroxyl fatty acids and α, ω-dicarboxyl fatty acids in core sediments of Gucheng Lake, Eastern China, Geochimica (in Chinese), 28(2): 183 (1999).

[4] Benson, L.V., Burdett, J.W., Kashgarian, M., et al. (1996). Climatic and hydrological oscillations in the Ownes Lake Basin, and adjacent Sierra Nevada, California, *Science* 274: 746-749.

[5] Meyers, P.A. & Ishiwatari, R. (1995). Organic matter accumulations records in lake sediments. In: Lerman, A., Imboden, D., Gat, J. (Eds.), Physics and Chemistry of Lakes, Springer-Verlag, Berlin, pp. 279-328.

[6] Benson, L., Burdett, L., Kashgarian, S., et al. (1997). Synchronous climate change in the Northern Hemisphere during the last glacial termination, *Nature* 388: 263-265.

[7] Gasse, F., Fontes, A.M., Fort, J.C., et al. (1991). 13, 000-year climate record from western Tibet, *Nature* 353: 742-745.

[8] Mayer, B. & Schwark, L. (1999). A 15,000-year stable isotope record from sediments of Lake Steisslingen, Southwest Germany, *Chemical Geology;* 161(1-3): 315-337.

[9] Wang, R.L., Scarpitta, S.C., Zhang, S.C., et al. (2002). Later Pleistocene/Holocene Climate Conditions of Qinghai-Xizhang Plateau (Tibet) Based on Carbon and Oxygen Stable Isotopes of Zabuye Lake Sediments, *Earth and Planetary Science Letters* 203(1): 461-477.

[10] An, Z.S., Wu, X.H. & Wang, P.X. (1991). Paleomonsoon of China for the last 130ka, *Science in China* (B) 10-11: 1076-1081.

[11] An, Z., Kutzbach, J.E., Prell W.L., et al. (2001). Evolution of Asian monsoons and phased uplift of the Himalaya? Tibetan plateau since Late Miocene times, *Nature* 411, 62 - 66.

[12] Hong, Y.T., Hong B., Lin, Q.H., et al. (2005). Inverse phase oscillations between the East Asian and Indian Ocean summer monsoons during the last 12 000 years and paleo-El Niño EPSL 231(3-4): 337-346.

[13] Wang, J.; Chen, X.; Zhu, X.; et al. (2001). Taihu Lake, lower Yangtze drainage basin: evolution, sedimentation rate and the sea level, *Geomorphology* 41; (2-3): 183-193.

[14] Wang, R.L., Brassell, S.C., Scarpitta, S.C., et al. (2004). Steroids in sediments from Zabuye Salt Lake, western Tibet: Diagenetic, climatic or ecological signals? *Organic Geochemistry* 35(2): 157-168.

[15] Kemp, A. (2000). Geology: Probing the memory of mud, *Nature* 406, 951 - 953.

[16] Hong, S, Candelone, J., Turetta, C., et al. (1996). Changes in natural lead, copper, zinc and cadmium concentrations in central Greenland ice from 8250 to 149,100 years ago: their association with climatic changes and resultant variations of dominant source contributions, *Earth and Planetary Science Letters,* 143: 233-244.

[17] Revel-Rolland, M., De Deckker, P., Delmonte, B., et al. (2006). Eastern Australia: A possible source of dust in East Antarctica interglacial ice, Earth and Planetary Science Letters, 249(1-2): 1-13.

[18] Nriagu, J.O. (1979). Global inventory of natural and anthropogenic emissions of trace metals to the atmosphere. *Nature* 279: 409-411.

[19] Nriagu, J.O. (1989). A global assessment of natural sources of atmospheric trace metals. *Nature* 338: 47–49.

[20] COHMAP Members (1988). Climatic changes of the last 18 000 years: observations and model simulations. *Science* 241:1043-1052.

[21] Zhou, W., Head, M.J. An, Z., et al. (2001). Terrestrial evidence for a spatial structure of tropical-polar interconnections during the Younger Dryas episode, *Earth and Planetary Science Letters;* 191: 231-239.

[22] Sigman, D.M. & Boyle, E.A. (2000). Glacial/interglacial variations in atmospheric carbon dioxide, *Nature;* 407: 859 - 869.

[23] Ruan J. & Wong M.H. (2001). Accumulation of Fluoride and Aluminium Related to Different Varieties of Tea Plant, *Environmental Geochemistry and Health,* 23(1): 53-63.

[24] Azovsky, A.I., Saburova, M.A., Chertoprood E.S., et al. (2005). Selective feeding of littoral harpacticoids on diatom algae: hungry gourmands? *Marine Biology,* 148(2): 327-337.

[25] Heusser, C.J., Heusser, L.E. & Peteet, D.M. (1985). Late-Quaternary climatic change on the American North Pacific Coast, *Nature;* 315: 485 - 487.

[26] Vincent, E., Killingley, J.S. & Berger, W.H. (1981). Stable isotope composition of benthic foraminifera from the equatorial Pacific, *Nature;* 289: 639 - 643.

[27] Morrill, C., Overpeck, J.T., Cole, J.E., et al. (2006). Holocene variations in the Asian monsoon inferred from the geochemistry of lake sediments in central Tibet, *Quaternary Research,* 65(2): 232-243.

[28] Peng, Y., Xiao, J., Nakamura, T., et al. (2005). Holocene East Asian monsoonal precipitation pattern revealed by grain-size distribution of core sediments of Daihai Lake in Inner Mongolia of north-central China, *Earth and Planetary Science Letters,* 233(3-4), 467-479.

[29] Verschuren, D., Laird, K.R., & Cumming, B.F. (2000). Rainfall and drought in equatorial east Africa during the past 1,100 years, *Nature* 403: 410-414.

[30] Stokes, S., Thomas, D.S.G. & Washington, R. (1997). Multiple episodes of aridity in southern Africa since the last interglacial period, *Nature* 388, 154 - 158.

[31] Guo, Z.T. Ruddiman, W.F., Hao, Q.Z., et al. (2002). Onset of Asian desertification by 22 Myr ago inferred from loess deposits in China, *Nature;* 416, 159 - 163.

[32] Hong, Y.T., Wang, Z.G., Jiang, H.B., et al. (2001). A 6000-year record of changes in drought and precipitation in northeastern China based on a $\delta^{13}C$ time series from peat cellulose, *Earth and Planetary Science Letters,* 185(1-2): 111-119.

[33] Dean, W.E., Bradbury, J.P., Roger Y., et al. (1984). The variability of Holocene climate change: evidence from varved lake sediments, *Science;* 226: 1191-11947.

Chapter 7

BIOGEOCHEMICAL EVALUATION OF SOIL COVERS FOR BASE METAL TAILINGS, AG-PB-ZN CANNINGTON MINE, AUSTRALIA

Benjamin S. Gilfedder[a] *and Bernd G. Lottermoser*[*]
School of Earth and Environmental Sciences, James Cook University, Townsville, Qld 4811, Australia

ABSTRACT

This study reports on the transfer of metals from soil covered tailings into native plants at the Cannington Ag-Pb-Zn mine in semi-arid northwest Queensland, Australia. A number of field trial plots were established over sulfidic metal-rich tailings in 2001. The plots differed in either soil depth (200, 500, 800 mm) or the combination of local soil and waste rock (1600 mm) used for the construction of the trialed capping strategy. In 2004, the plots were sampled for their cover materials, cover plants and tailings to evaluate the performance of the different cover designs. In all field trial plots, the roots and to a lesser degree the foliage of native plant species (*Triodia longiceps, Astrebla lappacea, Astrebla squarrosa, Iseilema membranaceum, Rhynchosia minima, Sclerolaena muricata*), growing on the soil covered tailings, display evidence of biological uptake of Ag, As, Cd, Pb, Sb and Zn, with values being up to one order of magnitude above background samples for the same species. The plants acquired their detected metal distributions from the tailings and mineralized waste rocks as evidenced by the penetration of plant roots through the entire soil cover to the top of the tailings or the mineralized waste rock layer. In general, the plant species growing on the soil covered tailings have bioconcentration factors (BCF, metal concentration ratio in plant roots to DTPA-extractable soil) and translocation factors (TF, metal concentration ratio of plant foliage to roots) for As, Cd, Cu and Zn and for Ag and Pb greater than one, respectively.

The trialled covers allow the translocation and accumulation of trace metals into the above-ground biomass of metal-tolerant native plants. Such processes may introduce metals

---

[a] Current address: Institute of Environmental Geochemistry, University of Heidelberg, Im Neuenheimer Feld 236, 69120 Heidelberg, Germany.
[*] Corresponding author: Telephone: +61-7-47816059; fax: +61-7-47814020. E-mail address: Bernd.Lottermoser@jcu.edu.au

and metalloids into surrounding ecosystems despite the waste remaining physically isolated. Hence, engineered dry covers of mine waste repositories need to consider the root penetration depth of native plants as well as the bioavailability of metals and their possible translocation and accumulation into the above-ground tissue of cover plants.

Keywords: Tailings; Plant uptake; Metals; Soil covers.

INTRODUCTION

Engineered dry covers of mine waste repositories have been developed to prevent migration of solid and dissolved contaminants from tailings storage facilities and waste rock dumps into the surrounding environment. The long-term performance of dry covers is strongly affected by the climate, material selection and availability, cost and construction practices, physical stability, volume change, soil evolution, ecological stability and vegetation growth (Wilson et al. 2003). Mine operators and government authorities install these technologies with the expectation that these covers will provide a long-term solution to waste isolation and will reduce the likelihood of interaction with the biosphere, acid generation, contaminant drainage and treatment costs. Moreover, mine sites should be able to be utilized in a productive manner following mine closure. Yet, there is little evidence of the proven long-term performance of engineered dry covers, despite the predicted design life of cover systems which extend into hundreds of years (Lottermoser 2007).

Vegetation is one key to the efficient functioning and sustainability of dry covers, partly because plants protect covers against erosion and transpire infiltrated water. Therefore, a solid understanding of vegetation performance on dry covers is crucial to the successful long-term establishment of vegetation over waste repositories (Bell and Menzies 2000). Nevertheless, the cover design of waste repositories still focuses on physical and engineering properties of the cover materials, with minimal consideration of vegetation growth needs and the effect of vegetation on cover performance in the long term. In particular, there is still little knowledge on the uptake of metals and metalloids into plants colonizing dry covers. This is despite the knowledge that salt-rich solutes rise in engineered waste covers in response to capillary action and evaporative suction forces, potentially threatening the integrity of cover systems (e.g. Elliott et al. 1997). For example, at the Rum Jungle uranium mine, the soil covers have been acidified and contaminated with Cu by the capillary rise of water from pyritic wastes underlying the soil capping (Menzies and Mulligan 2000). Plants take up metals from the capped tailings, introducing the contaminants into the surrounding environment and leading to vegetation dieback on the capped waste repository (Menzies and Mulligan 2000). As a result, the effectiveness offered by dry barrier-type covers can be compromised, even though the waste remains physically isolated.

The aims of this study were to evaluate dry cover systems for base metal tailings at the Cannington Ag-Pb-Zn mine in semi-arid Queensland, Australia. The study focuses on the uptake of metals and metalloids in plants colonizing dry covers, which had been installed over existing tailings and field trialed over a three-year period. Hence, this case study contributes to improving rehabilitation efforts of tailings repositories of base metal mine sites in semi-arid climates.

CANNINGTON MINE SITE

Physiography, Climate and Vegetation

The Cannington mine is located in northwest Queensland, Australia, approximately 200 km southeast of Mount Isa, at latitude 21°52'09" S, longitude 140°55'10" E (Figure 1). The mine lease is situated in the so-called "Channel Country" or "Mitchell Grass Plains". The Mitchell grasslands with their *Astrebla* and other grass species encompass 0.3 million km² of southeastern Queensland, the northern and central Northern Territory, the east Kimberley region of Western Australia, and the northern regions of South Australia.

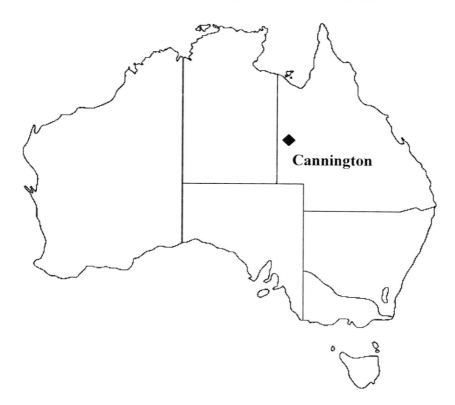

Figure 1: Location of the Cannington Ag-Pb-Zn mine, Australia.

The Cannington mine is located on the eastern flood plain of the Hamilton River. Streams in the region are ephemeral; few permanent water holes exist in the dry season and, significant flows are generally of only a few days duration after major rainfall events. It is evident that these flash-flood type events are the major cause of erosion and sediment transport in the region.

The climate is semi-arid to sub-humid tropical, with a marked summer rainfall maximum. Rainfall is highly variable; the annual average rainfall is 474 mm and annual evaporation greatly exceeds rainfall (Cloncurry weather station; approx. 140 km to the north of Cannington; BOM, 2007). Average daily temperatures range from a winter minimum of 10.3 °C to a summer maximum of 38 °C. Average daily temperatures range from a winter minimum of 10 °C to a summer maximum of 38 °C. Consequently, vegetation is relatively sparse, with native grasses (e.g. *Astrebla*), spinifex species, and small trees (eucalypts, acacias). Larger trees such as river red gums occur along some of the major ephemeral streams. The main land use in the district is low-density grazing.

Geology and Soils

The Cannington Ag-Pb-Zn deposit is located within the Eastern succession of the Proterozoic Mount Isa Inlier. The Cannington ore bodies are hosted by a sequence of migmatitic, biotite-sillimanite-garnet bearing quartzofeldspathic gneisses, with minor amphibolites (Bodon 1998; Walters and Bailey 1998). The deposit has undergone considerable prograde and retrograde metamorphism, extensive metasomatism, and a number of brittle and ductile deformations. The treated ores consist of major quartz, minor amounts of carbonate, chlorite, fluorite, garnet, magnetite, pyroxmangite and pyroxenes, and traces of hornblende, apatite, biotite, epidote, feldspar, gahnite, graphite, ilvaite, montmorillonite, muscovite, olivine, pyrosmalite, sillimanite and talc (Walters and Bailey 1998). The dominant sulfide minerals are sphalerite and galena. Trace sulfide minerals include arsenopyrite, chalcopyrite, freibergite, loellingite, marcasite, pyrargyrite, pyrite, pyrrhotite and a series of Ag-bearing Ag and Pb-Sb-Bi sulfides (Bodon 1998; Walters and Bailey 1998; Williams and Smith 2003). Prior to mining, the deposit contained at least 43.8 Mt with 11.6 % Pb, 4.4 % Zn and 538 g/t Ag (Walters and Bailey 1998). Geochemical analyses also show elevated As, Cd, Cu, F and Sb concentrations in the mined ore lenses (Walters and Bailey 1998).

The Cannington deposit is concealed beneath 10 to 60 m of Cretaceous and Quaternary sediments. These sediments have weathered to produce a vertosol soil, containing variable amounts of swelling clays such as nontronite. The expanding nature of these clay soils has led to a so-called 'self-mulching' soil whereby the shrinking and swelling of the expandable clays produces surface cracks. Soils in the mine area have been classified as self-mulching black to brown cracking clays, red cracking clays and scalded non-cracking clays on grasslands and plains, and sandy alluvial soils on flood plains.

Mineral Processing and Tailings Disposal

Since 1997, the Cannington process plant has employed grinding, flotation and filtration methods to produce a Zn concentrate, a Ag-rich Pb concentrate and a waste tailings stream. Various chemicals are used in the flotation circuit, particularly lime, sodium metabisulfide, xanthates, sulfuric acid and aluminium sulfate, which enter the tailings stream. As such, the pH of leachate effluent discharged to the tailings is often below 3, whereas the primary tailings have a natural to slightly basic pH. Tailings are either mixed with cement and backfilled into mined out stopes or are deposited in the tailings storage facility (TSF).

The tailings disposal system consists of a ~130 ha tailings repository with a decant pond and three tailings cells (~0.16 km^2). The repository was formed by embedding the facility into the underlying Quaternary clay-rich sediments and was constructed as a 'turkey nest' facility. The dam walls are raised by the upstream method when extra storage capacity is required. The current design allows for approximately 8.6 million m^3 of tailings storage.

The tailings range from fine sand to silty clay size particles. Previous environmental assessments of the tailings (French 2001; EGi 2002) demonstrated that the sulfidic tailings are acid generating and possess elevated sulfide (1.2-4.2 wt%) and fluoride (2.8-3.1 wt%) contents as well as high metal and metalloid concentrations (As 0.1-0.5 wt%; Fe 18-27 wt%; Mn 1.9-3.3 wt%; Pb 0.8-1.5 wt%; Zn 0.7-1.4 wt%). Exposure to the atmosphere has led to sulfide oxidation and acid production in the tailings. This can be seen by the abundance of sulfate efflorescences on the tailings surface and the low pH of tailings surface waters (pH <3). The chemistry of tailings surface waters is influenced by the F-rich nature of the milled ore and consequently the F-rich effluent from the leach circuit (pH 2-3) (Torrisi 2001).

In the tailings cells, grey-colored unoxidized tailings are generally overlain by red-brown-yellow colored oxidized tailings, which have a variable thickness (~20-65 cm). Oxidized tailings also occur at depth, and oxidation fronts occur as horizontal as well as vertical boundaries between the two tailings types. Compact surficial hardpans have formed in the tailings cells by evaporative concentration, dehydration and possibly also as paleosurfaces. All hardpans are present within red-brown-yellow oxidized tailings at the top level (<60cm) of the tailings piles. Surficial weathering of the tailings has also led to the contemporary formation of mineral efflorescences at and immediately below the tailings surface. Local efflorescent salt developments on tailings are apparently transient and form after each rain event and particularly during hot dry weather following such events. They include phases such as gypsum, plumbojarosite, natrojarosite, anglesite, halite, and native sulfur as indicated by XRD studies.

MATERIALS AND METHODS

Site Description

Capping of the Cannington TSF is to be based on a dry cover, using local native plant species (EMOS 2002). Therefore, in 2001 a number of field trial plots were established to evaluate the performance of different cover designs. The plots were constructed inside tailings cell no 2, covering a total area of 35x15 m and consisting of six different plots (each 2x2 m), replicated three times (French 2001). The plots differed in either soil depth or the combination of local soil horizons used for the construction of the trialed capping strategy. Soils used for tailings capping include local A, B and C vertosol horizons. Compared to the A and B soil horizon materials, the C horizon soil is coarser grained (ie. higher proportion of sand) and contains pebble sized rock fragments and gypsum. Each plot was sown with a local seed mix and fertilized with an N-P-K fertilizer (dynamic lifter). The six capping strategies comprised:

1) exposed tailings (trial plot 1);
2) 20 cm of A horizon soil (trial plot 2);
3) 50 cm of A horizon soil (trial plot 3);
4) 20 cm of A horizon underlain by 30 cm of C horizon soil (trial plot 4);

5) 20 cm of A horizon, 30 cm of B horizon and 30 cm of C horizon soil (trial plot 5); and
6) 20 cm of A horizon soil, 30 cm of B horizon soil, 30 cm of C horizon soil, 30 cm of mineralized waste rock (capillary break), 50 cm of C horizon soil and geotextile membrane (trial plot 6).

Sampling

Fieldwork was conducted in 2004, with the collection of several different sample media (tailings, mineral efflorescences, cover soils and plants, background soils and plants). Trenches were dug with a backhoe into six of the cover plots through the soils and 30 cm into the tailings (max. excavation depth of 1.9 m). Samples of soil cover materials (n: 16), tailings (n: 7) and mineralized waste rock (n: 1) were representative grab samples from various depths and constituted several kilograms. Samples of vegetation (complete specimens, n: 14) were taken from the individual cover plots (trial plots 2 to 6). In addition, tailings samples (n: 12) and mineral efflorescences were obtained from five pits and three drill cores outside the field trial plots and inside tailings cell no 2 (max. depth of 5 m).

The uptake of metals and metalloids from soils and tailings by plants can only be established relative to uncontaminated background sites. Therefore, a selection of soil from each soil horizon (n: 9) and vegetation (n: 6) samples was collected from locations outside the TSF. Such sites included the soil stockpiles, a control site established by French (2001), and sites up to 7 km from the mine lease.

Methods of Analysis

Soil and tailings samples were dried at 60°C, and ground in a chrome-steel ringmill. The sample powders were subsequently digested in a hot $HF-HNO_3-HCl$ acid mixture to determine total element contents. Selected tailings samples were also analyzed for their total carbon (C_{total}) and organic carbon (C_{org}) concentrations using a Leco furnace, while carbonate carbon (C_{carb}) was calculated by difference between the Leco methods (Australian Laboratory Services, ALS, Brisbane).

The bioavailability of trace elements in soils and tailings can be evaluated using extraction techniques such as DTPA (e.g. Menzies and Mulligan 2000; Mbila and Thompson 2004). Background soils (n: 3), soil cover materials (n: 16), waste rock (n: 1) and tailings (n: 6) were partially extracted using the DTPA method outlined in Rayment and Higginson (1992). The procedure is designed to establish plant available metal concentrations and was conducted as follows: approximately 15 g sample powder as starting material; add 30 mL of DTPA solution (0.005 M DTPA buffered with 0.1 M triethanolamine, 12 M HCl and 0.01 M $CaCl_2$ to pH 7.3); shake end-over-end for 2 hours; centrifuge and analyse the supernatant. Also, the solubility of major and trace elements in selected Cannington tailings (n: 6) was evaluated using a kinetic leaching method (cf. Lin and Herbert 1997). The water soluble fraction of the tailings was extracted by leaching the tailings samples with distilled water for 18 h at room temperature while undergoing constant agitation. Upon conclusion of the experiment, the leachates' pH as well as the trace and major element concentrations were measured. The extraction procedure was applied: (a) to study the release of trace elements and acid from tailings into pore waters; (b) to mimic leaching by rain water; and (c) to evaluate the retention of trace elements within the tailings.

Figure 2: View of the various trial plots. (a) Trial plot 1 with exposed tailings. (b) Trial plot 2 (20 cm of A horizon soil) with 50 % vegetation cover of Curley Mitchell grass, spinifex and Small Flinders grass. (c) Trial plot 3 (50 cm of A horizon soil) with 90 % vegetation cover of Curley Mitchell grass, spinifex and Small Flinders grass. (d) Trial plot 4 (20 cm A horizon and 30 cm B horizon soil) with 95 % vegetation cover of spinifex. (e) Trial plot 5 (20 cm A horizon, 30 cm B horizon and 30 cm C horizon soil) with 90 % vegetation cover of Black Roly-poly, Curley Mitchell grass, and spinifex. (f) Trial plot 6 (20 cm A horizon, 30 cm B horizon and 30 cm C horizon soil, 30 cm mineralized waste rock, 50 cm C horizon soil, geotextile) with 75 % vegetation cover of Bull Mitchell grass.

Plant samples taken from the trial covers for biogeochemical investigations include spinifex (*Triodia longiceps*), Curley Mitchell grass (*Astrebla lappacea*), Bull Mitchell grass (*Astrebla squarrosa*), Small Flinders grass (*Iseilema membranaceum*), Rhynchosia (*Rhynchosia minima*), and Black Roly-poly (*Sclerolaena muricata*) (Figure 2). Following collection, roots were trimmed and the above-ground biomass and the roots were subsequently washed with tap and distilled water, dried, and ashed at 520 °C. Ashed vegetation samples were dissolved using a $HCl-H_2O_2$ mixture.

Soils and tailings were analyzed by inductively coupled plasma atomic emission spectrometry (ICP-AES) for their total Ag, As, Ca, Cd, Cu, Fe, K, Mg, Mn, Na, P, Pb, S, Sb and Zn contents (ALS, Brisbane). The DTPA-extracts of soils and tailings as well as the plant digests were analyzed by inductively coupled plasma-mass spectrometry (ICP-MS) for trace elements (Ag, As, Cd, Cu, Pb, Sb, Zn). Quality control/assurance of the data was obtained by analyzing sample replicates, blanks and international reference standard GXR-3 with each sample batch. All analyses were within <10 % of the given values.

Paste pH measurements were performed on powdered tailings to provide a preliminary evaluation of the tailings' acid generation potential. Soil pH measurements were performed on sieved soils according to Rayment and Higginson (1992). The mineral constituents in selected mineral efflorescences, tailings and soil samples were identified by X-ray diffraction using a Siemens D5000 X-ray diffractometer and semi-automated peak search-match (James Cook University).

RESULTS

Tailings

Tailings samples from the various sample locations display no major differences in bulk composition. In addition, oxidized and anoxidized tailings possess similar mean, median as well as minimum and maximum element values, with the exception of C_{carb}, Ca, Mg, K and Na contents being distinctly higher in the unoxidized tailings. The distinct enrichment is likely due to the leaching and associated loss of C_{carb}, Ca, Mg, K and Na during weathering and acidification of the surface tailings. The tailings samples have major (ie. >1 wt%) median concentrations of Ca, Fe, Mn, Pb and S, minor (ie. >0.1 wt%) As, K, Mg, Na, P and Zn, sub-minor (ie. >100 ppm) Cu and Sb, and traces (ie. <100 ppm) of Ag and Cd (Table 1). The element distributions are consistent with the occurrence of relatively abundant primary aluminosilicates (ie. K-feldpsar, plagioclase, chlorite, talc, biotite, garnet, amphibole, hedenbergite, fayalite), magnetite and secondary minerals (ie. gypsum, plumbojarosite, natrojarosite, anglesite, halite, szomolnokite, native sulfur) as indicated by XRD and petrographic studies. The presence of sulfides (ie. sphalerite and galena) is evident in XRD traces and supported by the known mineralogy of processed ores and the abundance of S, Pb and Zn in the analysed tailings (Table 1). Oxidized tailings have the same primary mineralogy as the non-oxidized tailings. Powder XRD analyses of oxidized tailings did not reveal any crystalline Fe oxide or hydroxide phases. Hence, the oxidized red-brown tailings are impregnated by non-crystalline HFO (ie. hydrous ferric oxide) phases.

The DTPA treatment of tailings dissolved major (ie. >100 mg/L) mean concentrations of Pb and Zn, minor (ie. 1-10 mg/L) Cd and Cu, traces (ie. 0.01-1 mg/L) of As, and non-detectable Ag and Sb. The extractant leached 1-10 % of the total Cd, Cu, Pb and Zn, and <1% of the total Ag, As and Sb. The tailings had elevated plant available metal and metalloid concentrations in the order of Pb > Zn > Cu, Cd > As, whereas the tailings had total element values of Pb > Zn > As > Sb > Cu > Ag > Cd (Table 1).

Partial leaching of elements from tailings using distilled water only dissolved minor percentages of Cd and Zn from tailings samples (1.6 and 2.1 %, respectively). Extraction percentages of Ag, As, Cu, Pb and Sb were less than 1 %. Overall, the simulated pore waters of tailings had low metal and metalloid concentrations in the order of Zn > Pb > Cd, As > Cu > Sb > Ag. Tailings leachates from the upper tailings pile (ie. <5 cm from the surface) are distinctly acid (pH 3.5) and contain the highest simulated pore-water metal concentrations. By contrast, tailings from lower unoxidized parts of the tailings pile (ie. >1.5 m from the surface) have a near neutral pore water pH and lower pore water metal concentrations.

Paste pH measurements of milled tailings confirm the fact that the waste is acid producing as wastes have a median paste pH value of 4.7 (Table 1). Hence, the oxidation and wetting of tailings lead to the oxidation of sulfides, release of sorbed hydrogen and metal ions, the hydrolysis of iron and associated acid production, and the dissolution of acid producing mineral efflorescences (e.g. jarosite type phases). Thus, the investigated Cannington tailings have a minor net acid producing potential (NAPP). On the other hand, the tailings have a low acid neutralisation capacity (ANC) due to insignificant carbonate contents (C_{carb} <0.01-0.43 wt%) and the abundance of quartz and aluminosilicates.

Table 1: Total element concentrations of Cannington tailings (n = 19)

Element	Minimum	Maximum	Arithmetic Mean	Median
Major elements (WT%)				
C_{carb}	<0.01	0.43	0.16	0.10
Ca	3.99	7.60	5.61	5.34
Fe_{total}	11.35	19.10	15.11	14.35
K	0.39	1.07	0.70	0.70
Mg	0.44	1.91	0.99	0.94
Mn	0.87	1.98	1.46	1.50
Na	0.20	0.55	0.33	0.30
P	0.19	0.36	0.28	0.28
Pb	0.73	2.10	1.37	1.36
S_{total}	1.10	2.90	1.89	1.80
Zn	0.13	1.45	0.53	0.39
Trace elements (MG/KG)				
AG	24	105	61.3	56
As	1050	3490	1832	1835
Cd	4.1	65.4	24.7	18.2
Cu	66	196	123	111
Sb	146	297	212	204
paste pH	3.9	5.8	4.7	4.7

Soils

Soils are composed of quartz and feldspars, with minor kaolinite and variable proportions of swelling clays (probably nontronite as observed in XRD scans). Samples taken from the A, B and C horizon soil stockpiles display very similar geochemical compositions, with the exception of Ca and S. Compared to A horizon soils, the B and C horizon soils possess a greater quantity of swelling clays and distinctly higher Ca and S values (reflecting the increasing abundance of calcareous nodules, limestone pebbles and gypsum). The median trace element contents of the background soils are similar to their estimated abundances in soils elsewhere (e.g. Smith and Huyck 1999) (Table 2). The DTPA treatment dissolved minor median concentrations (ie. 1-100 mg/L) of Cu and Pb, traces (ie. 0.01-1 mg/L) of Cd and Zn, and non-detectable Ag, As and Sb. The extractant leached >10 % of the total Pb and 1-10 % of the total Cd, Cu and Zn.

Table 2: Total element concentrations (n = 9) and DTPA-extracts (n=3) of background soils

Element	Minimum	Maximum	Arithmetic Mean	Median
Major elements (WT%)				
Ca	0.10	1.54	0.80	0.63
Fe_{total}	3.01	4.01	3.66	3.75
K	0.97	1.43	1.23	1.25
Mg	0.35	0.62	0.49	0.46
Mn	434	720	550	545
Na	0.56	0.89	0.77	0.82
S_{total}	0.01	0.97	0.33	0.12
Trace elements (MG/KG)				
AG	<0.5	<0.5	<0.5	<0.5
As	5	14	9.2	10
Cd	<0.5	<0.5	<0.5	<0.5
Cu	20	34	25.3	23
P	200	300	250	240
Pb	12	20	17	17
Sb	<5	<5	<5	<5
Zn	47	61	55.9	57
pH	5.95	8.86	7.27	7.11
DTPA extractable trace elements (mg/kg)				
As	<0.02	<0.02	<0.02	<0.02
Cd	0.014	0.044	0.025	0.018
Cu	1	1.17	1.08	1.07
Mn	10.7	32.2	21.6	21.8
Pb	1.9	2.77	2.29	2.19
S	8	1258	774	1058
Zn	0.68	0.68	0.59	0.68

Soil Covers

Inspection of the trial plots 2 to 6 revealed that desiccation, cracking and compaction of soil covers have occurred since their installation in 2001. Soil cover materials emplaced above the tailings exhibit plot specific element profiles (Table 3). In plots with thicker soil covers (50-1600 mm), the lowest concentrations of trace elements (in particular Pb and Zn) are present in the intermediate layers of the cover sequences (trial plots 3 to 6, Table 3). The mid cover materials are not in physical contact with the tailings and are not exposed to the surface, and they possess median trace element concentrations which are not significantly different from background soils. By contrast, soil horizons directly above the tailings and at the surface of these plots have distinctly higher trace element contents (trial plots 3 to 6, Table 3). In particular, the trace elements Pb and Zn show a pronounced increase in concentration towards the top and bottom of these thicker cover profiles. Cover plot 6 is slightly different from the soil-only profiles, as the metal concentrations in the soil horizon above the waste rock layer were similar to background levels (Table 3). This suggests that the capillary break of mineralized waste rock was adequate at suppressing the vertical movement of metals from the waste rock material itself.

Table 3: Total element concentrations (mg/kg) and soil pH values of cover profiles

Sample location	Depth (cm) from surface	pH	Ag	As	Cd	Cu	Pb	Sb	Zn
Trial plot 2	0-10	7.85	<0.5	13	<0.5	28	135	<5	126
	10-25	8.25	1.4	17	1.8	27	508	<5	300
	25-33	7.3	4.9	414	4.1	57	1500	7	769
	>33 (tailings)	5.6	55.3	1910	13.4	108	14600	211	3260
Trial plot 3	0-20	7.3	<0.5	20	<0.5	21	149	<5	92
	20-40	8.2	<0.5	8	<0.5	27	20	<5	55
	40-50	7.45	<0.5	9	1	23	101	<5	158
	>50 (tailings)	4.95	46.8	1640	12.6	97	13000	181	2810
Trial plot 4	0-20	8.1	<0.5	<5	<0.5	24	41	<5	69
	20-30	7.8	<0.5	8	<0.5	27	16	<5	55
	30-50	7.25	<0.5	16	0.6	22	48	<5	147
	>50 (tailings)	7.15	54.2	1970	20.3	111	15600	206	3920
Trial plot 5	0-20	8.13	<0.5	15	<0.5	28	28	<5	56
	20-57	8.0	<0.5	9	<0.5	31	22	<5	60
	57-76	6.9	<0.5	6	1	18	35	<5	196
	76-89 (tailings)	4.95	47.6	1755	10.6	98	14200	182	2590
	>89 (tailings)	6.92	66.5	1835	35.6	140	19400	204	6550
Trial plot 6	0-20	7.9	<0.5	8	<0.5	29	51	<5	68
	20-50	7.69	<0.5	9	<0.5	28	17	<5	56
	50-68	7.31	<0.5	11	<0.5	22	22	<5	47
	68-100 (waste rocks)	6.5	88.7	367	45.6	386	21000	95	9300
	100-120	7.07	<0.5	7	2	26	40	<5	321
	>120 (tailings)	4.36	52.7	1925	10.4	152	13000	200	2920

The DTPA treatment of the soil cover materials dissolved minor median concentrations (ie. 1-100 mg/L) of Pb and Zn, traces (ie. 0.01-1 mg/L) of As, Cd and Cu, and non-detectable

Ag and Sb. The upper and lower soil horizons thereby possess higher plant available metal (Cd, Cu, Pb, Zn) concentrations than the intermediate soil profile zones (Table 4). In comparison to these intermediate zones and the background soils, the upper and lower soil horizons have distinctly higher plant available metal concentrations (Table 4). Thus, the upper and lower horizons of soil covers overlying Cannington tailings have gained total as well as plant available Cd, Cu, Pb and Zn since their installation on the TSF. While the distinctly elevated metal concentrations at the top of the soil profiles are most likely the result of tailings dust contamination from the adjacent tailings cells, the metal enrichment of the lower horizons in the soil covers has to be due to the capillary rise of metal-rich pore waters from the tailings. Again, the waste rock layer appeared to limit the capillary rise of metals from the tailings into the overlying soil.

Table 4: DTPA-extractable element concentrations (mg/l) of cover profiles

Sample location	Depth (cm) from surface	As	Cd	Cu	Pb	Zn
Trial plot 2	0-10	0.04	0.04	2.53	50	13.9
	10-25	<0.01	0.12	1.9	45	11.2
	25-33	0.06	0.72	1.49	316	95
	>33 (tailings)	0.08	1.50	0.85	846	222
Trial plot 3	0-20	0.06	0.13	0.99	77.6	10.9
	20-40	<0.01	0.02	1.12	8.38	1.78
	40-50	0.06	0.52	0.92	121	53.8
	>50 (tailings)	0.04	2.29	5.08	260	412
Trial plot 4	0-20	<0.01	0.06	1.03	23	7.02
	20-30	<0.01	0.01	0.87	6.04	1.96
	30-50	<0.01	0.23	0.67	10.7	24.4
	>50 (tailings)	0.08	3.04	0.83	1368	164
Trial plot 5	0-20	<0.01	0.01	1.14	5.44	1.06
	20-57	<0.01	0.01	0.61	1.96	0.82
	57-76	<0.01	0.74	0.23	8.42	122
	76-89 (tailings)	0.08	2.24	2.48	644	218
	>89 (tailings)	0.14	1.75	0.22	1338	134
Trial plot 6	0-20	<0.01	0.02	1.32	18.5	2.3
	20-50	<0.01	0.01	0.35	2.24	0.58
	50-68	<0.01	0.01	0.25	2.47	0.5
	68-100 (waste rocks)	<0.01	2.75	1.70	1572	175
	100-120	0.04	0.83	0.42	130	69
	>120 (tailings)	0.04	1.33	4.96	368	167

Plants

Several plant species colonised the trial plots including spinifex (*Triodia longiceps*), Curley Mitchell grass (*Astrebla lappacea*), Bull Mitchell grass (*Astrebla squarrosa*), Small Flinders grass (*Iseilema membranaceum*), Rhynchosia (*Rhynchosia minima*), and Black Rolypoly (*Sclerolaena muricata*). Inspection of the trial plots revealed that plant roots had penetrated through the entire soil cover to the top of the tailings (trial plots 2 to 5) or the mineralized waste rock layer (trial plot 6) (Figure 3). Also, many plant roots had extended laterally along the tailings surface and were encrusted with secondary salts.

Figure 3: View of pit dug into trial plot 5 (800 mm soil cover). (a) Plant roots are present throughout the soil cover profile. Soil cracking is evident at the surface (top right-hand side of the photograph). (b) Close-up photograph of the tailings-soil cover interface, showing abundant plant roots.

Vegetation samples were also taken outside the TSF and are regarded as representing "background" material. All background materials tend to contain significantly lower values of metals and metalloids in their foliage and roots than those of plant materials from the rehabilitation plots (Table 5). Vegetation samples taken outside the mine site occur on soil containing low metal and metalloid contents. Vegetation samples from the rehabilitation plots grow on substrates with higher metal and metalloid values and the plant samples clearly reflect the higher concentrations. A comparison between the same species growing on "background" versus rehabilitation plots shows that particularly the roots of plant species from the trial plots tend to contain higher metal (Ag, Cd, Cu, Pb, Zn) and metalloid (As, Sb) contents (Table 5).

Table 5: Total element concentrations (in mg/kg dry weight) of plant samples from the field trial plots and background sites

Sample types	Ag	As	Cd	Cu	Pb	Sb	Zn
Roots							
Spinifex (*Triodia longiceps*), n: 3	0.12	2.47	1.31	9.84	25.4	0.16	295
Spinifex (*Triodia longiceps*), background, n: 1	0.03	0.17	0.51	15.52	7.02	0.03	17
Curly Mitchell grass (*Astrebla lappacea*), n: 4	0.08	0.95	0.67	17.5	17.6	0.10	84.8
Curly Mitchell grass (*Astrebla lappacea*), background, n: 1	0.03	0.15	0.18	5.47	2.95	0.03	11
Bull Mitchell grass (*Astrebla squarrosa*), n: 1	0.06	0.65	1.2	12.56	5.89	0.16	38
Bull Mitchell grass (*Astrebla squarrosa*), background, n: 1	0.06	0.25	0.14	5.6	2.6	0.16	14
Small Flinders grass (*Iseilema membranaceum*), n: 3	0.47	7.16	2.43	10.1	57.1	0.58	233
Small Flinders grass (*Iseilema membranaceum*), background, n: 1	0.07	0.93	0.35	8.32	26.8	0.14	37
Black Roly-poly (*Sclerolaena muricata*), n: 2	0.14	0.24	3.50	7.00	7.12	0.06	274
Black Roly-poly (*Sclerolaena muricata*), background, n: 1	0.05	0.1	0.14	7.13	7.13	0.02	13
Rhynchosia (*Rhynchosia minima*), n: 1	0.05	0.65	6.69	8.6	30.31	0.08	364
Rhynchosia (*Rhynchosia minima*), background, n: 1	0.02	<0.001	0.08	1.36	1.36	0.02	31
Foliage							
Spinifex (*Triodia longiceps*), n: 3	0.12	1.08	1.53	3.22	37.5	0.12	233
Spinifex (*Triodia longiceps*), background, n: 1	0.14	0.86	0.40	1.91	24	0.24	46
Curly Mitchell grass (*Astrebla lappacea*), n: 4	0.28	3.84	0.84	21.8	128	0.37	150
Curly Mitchell grass (*Astrebla lappacea*), background, n: 1	0.35	0.80	0.24	8.92	61.9	0.19	29
Bull Mitchell grass (*Astrebla squarrosa*), n: 1	0.17	1.39	0.20	9.49	48.1	0.23	32
Bull Mitchell grass (*Astrebla squarrosa*), background, n: 1	0.04	<0.01	0.11	2.74	3.31	0.02	5
Small Flinders grass (*Iseilema membranaceum*), n: 3	0.32	1.95	1.67	15.6	98.4	0.24	283
Small Flinders grass (*Iseilema membranaceum*), background, n: 1	0.34	0.78	0.34	8.58	86.1	0.18	77
Black Rolypoly (*Sclerolaena muricata*), n: 2	0.16	0.60	4.53	5.91	27.5	0.19	366
Black Rolypoly (*Sclerolaena muricata*), background, n: 1	0.22	0.30	0.43	42.4	42.4	0.21	53
Rhynchosia (*Rhynchosia minima*), n: 1	0.11	0.98	3.28	4.35	32.2	0.16	281
Rhynchosia (*Rhynchosia minima*), background, n: 1	0.11	0.35	0.20	27.1	27.1	0.11	32

DISCUSSION

Covers in semi-arid environments rely on maintaining a healthy vegetative cover as transpiration is the primary means of removing moisture stored in the cover profile and limiting erosion over the confining layer. While numerous studies have investigated the evaporative and transporative properties of store-and-release covers, there is very little known about the ability of plants to take up and translocation metals into their foliage. This is unexpected given the importance of vegetation to these cover types and the ability of metals and metalloids to accumulate in plants. In such cases, metals may be introduced into surrounding food webs (by grazing on the colonizing vegetation) despite the metal-rich waste material remaining physically isolated.

A plant's ability to accumulate metals from soils can be quantified using the bioconcentration factor (BCF) (Yoon et al. 2006). In this study, the BCF is defined as the ratio of metal concentration in the roots to the DTPA-extractable metal concentrations in the soils. Moreover, a plant's ability to translocate metals from the roots to the foliage can be quantified using the translocation factor (TF), which is defined as the ratio of metal concentration in the foliage to that in the roots. Metal tolerant plant species with high BCF and low TF values can be used for the rehabilitation of metalliferous ground (Sursala et al. 2002; Yoon et al. 2006). Such plants immobilize metals through adsorption and accumulation by roots, adsorption onto roots, or precipitation within the rhizosphere. However, if a metal is taken up by a plant and accumulated into the foliage, the plant will exhibit TF values greater than one. Such plants are not suitable for mine site rehabilitation as they allow the accumulation and translocation of metals from mineralized wastes or metal-bearing porewaters into plant roots and then into the above-ground biomass.

At Cannington, the six plant species collected from the trial plots have BCF values greater than one for As, Cd, Cu and Zn (Table 6). The TF values for all plants are also greater than one for Ag and Pb (Table 6). These elevated BCF and TF values reflect the distinct accumulation and translocation of bioavailable trace metals from the plants' roots to their foliage. The grass species *Astrebla lappacea* and *Astrebla squarrosa* growing on the trial plots have the highest TF value for Pb of all analyzed plant samples. The two grass species clearly have the ability to translocate trace metals when growing on soil covered tailings, particularly Pb and to a lesser degree of Ag, As, Cd, Sb and Zn (Table 6). Therefore, the trialed soil covers allow the accumulation and transfer of metals and metalloids from mineralized waste into the above-ground biomass of native plants.

In selecting plant species for tailings revegetation, numerous criteria such as drought tolerance, stabilization ability and rooting depth need to be taken into consideration (Mulligan 1998). At Cannington, the roots of native grass (*Triodia longiceps*, *Astrebla lappacea*, *Astrebla squarrosa*, *Iseilema membranaceum*), creeper (*Rhynchosia minima*) and herb (*Sclerolaena muricata*) species are capable of penetrating significant soil cover depth. Moreover, these plants have the propensity to accumulate metals from the covered waste materials. Therefore, any modified cover design using native plants should explore chemical amendments (e.g. lime or phosphate) because such materials may reduce the bioavailability of trace metals, potentially leading to lower metal concentrations as well as higher BCF and lower TF values in native plants. The case study has demonstrated that engineered dry covers of mine waste repositories need to consider the bioavailability of metals and their possible translocation and accumulation into the above-ground tissue of cover plants.

Table 6: Bioconcentration and translocation factors for metals and metalloids in plant samples from the field trial plots. The bioconcentration factors for Ag and Sb could not be determined as the DTPA-extracts revealed non-detectable Ag and Sb concentrations

Bioconcentration factor (BCF)	Ag	As	Cd	Cu	Pb	Sb	Zn
Spinifex (*Triodia longiceps*), n: 3		54.17	1.94	26.30	0.44		3.60
Curly Mitchell grass (*Astrebla lappacea*), n: 4		31.17	1.97	20.04	1.08		1.94
Bull Mitchell grass (*Astrebla squarrosa*), n: 1		16.25	1.43	29.62	0.05		0.55
Small Flinders grass (*Iseilema membranaceum*), n: 3		84.97	3.07	15.42	0.22		2.11
Black Rolypoly (*Sclerolaena muricata*), n: 2		21.00	4.43	22.91	0.39		2.38
Rhynchosia (*Rhynchosia minima*), n: 1		65.00	28.84	12.68	2.81		14.92
Translocation factor (TF)	**Ag**	**As**	**Cd**	**Cu**	**Pb**	**Sb**	**Zn**
Spinifex (*Triodia longiceps*), n: 3	1.27	0.84	1.44	0.40	1.36	0.89	1.00
Curly Mitchell grass (*Astrebla lappacea*), n: 4	4.87	5.66	1.34	0.98	7.53	4.13	1.85
Bull Mitchell grass (*Astrebla squarrosa*), n: 1	2.83	2.14	0.17	0.76	8.16	1.44	0.84
Small Flinders grass (*Iseilema membranaceum*), n: 3	1.16	0.42	0.57	1.71	1.79	0.56	1.32
Black Roly-poly (*Sclerolaena muricata*), n: 2	1.35	3.89	1.29	0.84	3.97	3.53	7.03
Rhynchosia (*Rhynchosia minima*), n: 1	2.20	1.51	0.49	0.51	1.06	2.00	0.77

CONCLUSION

This study was conducted to evaluate the performance of trialed dry covers and the uptake of metals and metalloids by native plants growing on soil covered tailings at the Cannington Ag-Pb-Zn mine. Colonizing plant species included *Triodia longiceps, Astrebla lappacea, Astrebla squarrosa, Iseilema membranaceum, Rhynchosia minima,* and *Sclerolaena muricata*. The roots of these plants had penetrated through the entire soil cover to the top of the buried tailings or the mineralized waste rock layer. Therefore, the plants acquired the detected metal distributions from the metalliferous tailings and mineralized waste rocks. Plant species growing on soil covered tailings have bioconcentration factors (BCF, metal concentration ratio in plant roots to DTPA-extractable soil) and translocation factors (TF, metal concentration ratio of plant foliage to roots) for As, Cd, Cu and Zn and for Ag and Pb greater than one, respectively.

The trialed covers have allowed the translocation and accumulation of metals in cover plants. Hence, dry cover designs of tailings storage facilities and waste rock dumps need to consider the root penetration depth of metal-tolerant cover plants as well as the bioavailability of metals and their possible translocation and accumulation into the above-ground tissue of cover plants.

ACKNOWLEDGEMENTS

Support for this project was given by the Australian Research Council, James Cook University and BHPBilliton Cannington.

REFERENCES

Bell, L.C. & Menzies, N.W. (2000). In *Proceedings of fourth Australian workshop on acid mine drainage;* Editors Grundon, N.J., Bell, L.C. Australian Centre for Mining Environmental Research, Kenmore, Queensland, Australia, pp 171-177.

Bodon, S.B. (1998). *Econ Geol.* 93, 1463-1488.

BOM (Bureau of Meteorology) (2007). Climate averages for Cloncurry, Queensland. http://www.bom.gov.au/climate/averages/tables/cw_029141.shtml.

EGi (Environmental Geochemistry International) Cannington's north block extension project: environmental geochemical assessment of waste rock and tailings. *Environmental Geochemistry International*, Sydney, 2000, unpubl. Report to BHPBilliton.

Elliott, L.C.M.; Davison, G.J.; Liangxue, L.; et al. (1997). In *Fourth international convention on acid rock drainage*. Vancouver, pp 1213-1228.

EMOS (2002). Environmental management overview strategy – Cannington mine. BHPBilliton, unpub., 2002

French, S. (2001). Unpubl. *Applied Science thesis*, University of Queensland, Brisbane.

Lin, Z. & Herbert, R.B. (1997). In *Fourth international conference on tailings and mine waste '97*; Editor Nelson, J. D. Balkema, Rotterdam, pp 237-246.

Lottermoser, B.G. (2007). *Mine wastes: characterization, treatment, environmental impacts*; Springer: Berlin Heidelberg New York 304pp.

Mbila, M.O. & Thompson, M.L. (2004). *J Environ Qual.* 33, 553-539.

Menzies, N.W. & Mulligan, D.R. (2004). *J Environ Qual.* 29, 437-442.

Mulligan, D. (1998). In *Proceedings of workshop on future directions in tailings environmental management*; Editors Asher, C.J., Bell, L.C. Australian Centre for Mining Environmental Research, Kenmore, Queensland, Australia, pp 159-166.

Rayment, G.E. & Higginson, F.R. (1992). *Australian laboratory handbook of soil and water chemical methods*. Inkata Press, Port Melbourne, 330pp.

Smith K.S. & Huyck H.L.O. (1999). In The environmental geochemistry of mineral deposits. Part A: Processes, techniques and health issues; Editors Plumlee, G. S.; Logsdon, M. J. Society of Economic Geologists, *Reviews in Economic Geology*, Volume 6A, pp 29-70.

Sursala, S.; Medina, V.F. & McCutcheon, S.C. (2002). *Ecol Eng.* 18, 647-658.

Torrisi, C. (2001). *Minerals Eng.* 14, 1637-1648.

Walters, S. & Bailey, A. (1998). *Econ Geol.* 93, 1307-1329.

Williams, P.J. & Smith, M. J. (2003). *Geochem Explor Environ Anal.* 3, 245-261.

Wilson, G.W.; Williams, D.J. & Rykaart, E.M. (2003). In Sixth international conference on acid rock drainage. Australasian Institute of Mining & Metallurgy, Melbourne, pp 445-461.

Yoon, J.; Cao, X.; Zhou, Q.; et al. (2006). *Q. Sci Total Environ.* 368, 456-464.

Chapter 8

THE GEOCHEMICAL CHARACTERISTICS OF RARE EARTH ELEMENTS IN GRANITIC LATERITES IN HAINAN ISLAND, CHINA*

Liu Qiang,[1] Bi Hua,[2] Wang Xueping,[2] Wang Minying,[1] Zhao Zhizhong,[2] Yang Yuangen[3] and Zhu Weihuang[3]

1 Department of Biology, Hainan Normal University, Haikou 571158, China
2 Department of Resources, Environment and Tourism, Hainan Normal University, Haikou 571158, China;
3 Institute of Geochemistry, CAS, Guiyang 550002, China

ABSTRACT

In granitic laterites across Hainan Island the average grosses of REEs, light REEs and available REEs are 459.25, 386.99 and 199.06 µg/g respectively. The above three values are higher than that of the crust 186 µg/g, but the average gross of heavy REEs is 72.26 µg/g. Light REEs are enriched. Available REEs are high and have significantly positive correlation with gross of REEs, accounting for the total REEs in the range of 17.99%-80.25% and with mean value of 43.35%. It is also related to high temperature and humidity, strong chemical weathering and acid to weak acid soil environment in tropical Hainan Island. Across Hainan Island the average grosses of REEs, light REEs and available REEs decline along with low mountain-hilly land → central mountain → tableland-coastal plain on the horizontal spatial dimension, while the average grosses of heavy REEs decline along with central mountain →lower mountain-hilly land→ tableland-coastal plain. The average grosses of REEs, and Light REEs decline from bottom to top in laterite (Horizon C → Horizon B → Horizon A), while heavy REEs and available REEs decline from bottom to top (Horizon C → Horizon B → Horizon A) gradually on the vertical spatial dimension. However in tableland areas heavy REEs and available REEs have similar vertical change patterns as REEs and Light REEs. All laterites show negative Eu-anomaly to weak-medium extent. It shows relative enrichment of Ce and positive Ce-anomaly in top layer (Horizon A) of granitic laterites in all geomorphic

---

* Project supported by the National Science Foundation of China (40061003)

units and in lower altitude tableland-coastal plain, but on the vertical spatial dimension it shows the change from no anomaly to weak negative anomaly with the depth of the soils.

Before our project being carried out, no report about systematic study on rare earth elements (i.e. REEs) in laterites in Hainan Island has been found [1]. Over 5 years' study, we surveyed the typical profiles of laterites on different mother materials, in different structural landform and with different maturity, collected, processed and analyzed many samples in a systematic way. We have conducted REE fertilizing experiments both in potted plants and in field growing plants. In this article we focus on the geochemical characteristics of REE in granitic laterites in Hainan Island.

Keywords: Granitic Laterites; Rare Earth Elements; Geochemistry; Hainan Island

1 BASIC GEOMORPHOLOGIC CHARACTERISTICS IN HAINAN ISLAND

Hainan Island (18°10′~20°10′N, 108°37′~111°03′E), located in South China Sea, is the second largest island after Taiwan Island in China. It covers a land area of 33,920 km^2, accounting for 42.5% of China's total tropical land and being well know as China's largest tropical treasure island. It is in the shape of an ellipse which trends to NE-SW, whose central area is higher than the surrounding areas. From the central area to the surrounding areas, the landform is gradually descends from mountains, hills, tableland-terrace, coastal plain, and littoral sandbank to bottomland, which forms the structure of anisomerous multilevel vertical layers-annularity (Figure 1 and Table 1).

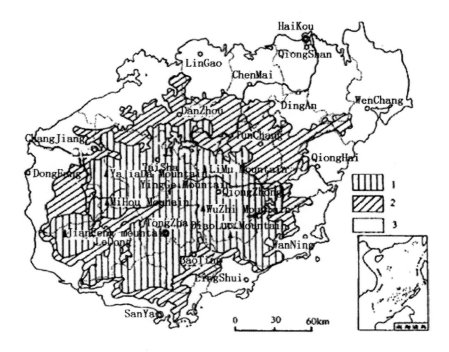

1, middle mountain; 2, low-mountain and hill; 3, tableland plain

Figure 1: Geomorphological structure of Hainan Island [2-7].

Table1: Main physiognomy types in Hainan Island

Physiognomy types		Altitude / m	Area/ km²	Percentage of the land area in Hainan Island (%)
Mountains	Middle mountains	>1000	6067.60	17.9
	Low mountains	500-1000	2555.45	7.5
	Hills	100-500	4497.71	13.3
Tableland-terrace	Tableland	<100	11052.40	32.6
	Terrace		5724.36	16.9
Plain			3808.85	11.2
Anthropogenic landforms			214.16	0.6
total			33920.53	100
mudflat		-5 ~ -10	2330	

Note: data come from literature [2-6].

2 GENERAL SITUATION ABOUT GRANITES AND GRANITIC LATERITES

There were granite magmatic intrusions in Jinning stage in Middle-Proterozoic era, Haixi stage in Late-Palaeozoic era, Indo-Chinese to Yanshan epoch in Mesozoin era, and it was gradually formed that the geomorphology of granitic mountains-hills and a few tablelands in central and southern Hainan Island, which accounts for 37% of the land area in Hainan Island, about 12,400 km²*. Because of the high temperature and moisture climates granites changed into granitic laterites after weathering and pedogenesis for a long time. It is the main type of laterites in Hainan Island.

3 MATERIALS AND METHODS

Laterite soil samples were collected from 2000 to 2004. 24 sampling sites located in the different geomorphic units in the whole Hainan Island, such as middle mountains, low mountains-hills and tableland-coastal plain. Soils were sampled from leached horizon (A), illuvium (B) and Chorizon (C) of fresh soil profiles. In the same sampling site 3 soil samples from the same horizon were collected and sufficiently mixed and quartered, and one portion was randomly selected from the four. After air drying, grinding, eliminating gravel and remnants of animals-plants, blending and passing through a 0.074 mm sieve, each sample of 20g was prepared for analysis.

The REE content of soil samples was determined by ICP-MS in Institute of Geochemistry, Chinese Academy of Sciences. The average REE gross were calculated by different horizons, i.e. leached horizon (A), illuvium (B) and Chorizon (C) in three geomorphic units (middle mountains, low-mountain and hills, and tableland plain).

---

* Hainan geologic team, geologic bureau, Guangdong province. Introduction to geologic map of Hainan Island (1:20,0000), 1977.

4 RESULTS AND DISCUSSION

The concentration of REE and related parameters of granitic laterites in Hainan Island are shown in Table 2. The geochemical characteristics of REE in granitic laterites in Hainan Island are discussed as follows.

4.1 Gross of REE

For the whole island the average gross of REE in granitic laterites is 459.25 μg/g, light REE is 386.99 μg/g, and soluble REE (i.e. available REE) is 199.06 μg/g. All the 3 values are higher than that of the crust 186 μg/g[8] . But the average gross of heavy REE is 72.26 μg/g.

4.2 REE Speciation

The ratio of average gross of light REE and heavy REE ($\sum Ce/\sum Y$) in the granitic laterites is 5.36 (>1), and the average ratio of standard values of lanthanum and ytterbium chondritic meteorite is 16.59 (>1), which indicates that granitic laterites enriched light REE and made a deficit of heavy REE. The curves of average gross of REE slant to the right (Figure 2).

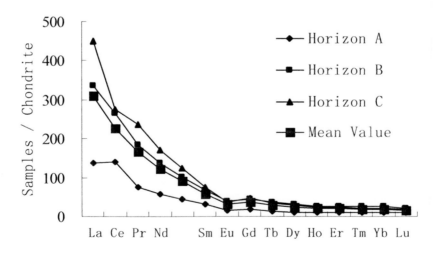

A, leached horizon; B, illuvium; C, chorizon

Figure 2 Distribution patterns of REE in granitic laterite profile in Hainan Island

Table 2: The REE Content (μg/g) and Relevant Parameters of Granitic Laterite in Hainan Island

Parameter	Horizon	Tableland plains A	B	C	average	Low-mountain hills A	B	C	average	Middle mountains A	B	C	average	Mean value A	B	C	average
REE content	La	20.93	59.16	127.07	69.05	63.01	142.61	178.25	127.95	45.18	111.86	114.80	90.61	43.04	104.54	140.04	95.87
	Ce	78.31	228.27	200.37	168.98	165.26	263.49	245.25	224.67	97.20	149.40	225.81	157.47	113.59	213.72	223.81	183.71
	Pr	3.59	9.47	24.14	12.40	14.40	33.40	35.65	27.82	9.90	24.59	27.29	20.59	9.30	22.49	29.02	20.27
	Nd	12.20	33.55	82.47	42.74	57.27	115.70	120.21	97.73	35.38	97.99	107.51	80.29	34.95	82.41	103.40	73.59
	Sm	2.18	4.74	11.07	6.00	9.33	16.72	14.68	13.58	6.24	17.90	19.15	14.43	5.91	13.12	14.96	11.33
	Eu	0.33	0.63	1.50	0.82	2.12	3.27	2.27	2.55	0.90	4.89	4.06	3.28	1.12	2.93	2.61	2.22
	Gd	2.57	4.23	7.99	4.93	6.58	11.80	9.92	9.43	4.59	19.01	17.68	13.76	4.58	11.68	11.86	9.37
	Tb	0.49	0.59	1.09	0.72	0.83	1.44	1.23	1.17	0.59	2.99	2.59	2.06	0.64	1.67	1.64	1.32
	Dy	3.58	3.67	5.91	4.38	4.30	7.47	6.01	5.92	3.25	18.47	14.65	12.12	3.71	9.87	8.85	7.48
	Ho	0.78	0.74	1.14	0.89	0.74	1.35	1.02	1.04	0.63	3.76	2.85	2.41	0.72	1.95	1.67	1.45
	Er	2.25	2.25	3.34	2.61	2.18	3.78	2.89	2.95	1.95	10.97	8.15	7.02	2.13	5.67	4.79	4.19
	Tm	0.35	0.34	0.49	0.39	0.32	0.52	0.39	0.41	0.34	1.55	1.13	1.00	0.33	0.80	0.67	0.60
	Yb	2.50	2.19	3.19	2.63	2.08	3.48	2.37	2.64	1.99	9.99	7.28	6.42	2.19	5.22	4.28	3.90
	Lu	0.35	0.30	0.45	0.37	0.30	0.46	0.32	0.36	0.28	1.37	1.02	0.89	0.31	0.71	0.60	0.54
	Y	22.17	21.34	35.75	26.42	21.42	41.55	30.42	31.13	17.46	118.94	81.67	72.69	20.35	60.61	49.28	43.41
	ΣREE	152.57	371.47	505.96	343.33	350.13	647.04	650.85	549.34	225.87	593.69	635.63	485.06	242.86	537.40	597.48	459.25
	ΣCe	117.54	335.83	446.61	299.99	311.38	575.20	596.31	494.29	194.79	406.63	498.62	366.68	207.90	439.22	513.85	386.99
	ΣY	35.03	35.64	59.35	43.34	38.75	71.84	54.55	55.05	31.08	187.06	137.01	118.39	34.96	98.18	83.64	72.26
	ΣCe/ΣY	3.36	9.42	7.53	6.92	8.04	7.95	10.93	8.98	6.27	2.17	3.64	3.10	5.95	4.47	6.14	5.36
Standard value	δEu	0.43	0.43	0.49	0.46	0.83	0.71	0.58	0.69	0.52	0.81	0.67	0.71	0.66	0.72	0.60	0.66
	δCe	2.17	2.32	0.87	1.39	1.32	0.92	0.74	0.91	1.11	0.69	0.97	0.88	1.37	1.06	0.85	1.00
	$(La/Yb)_N$	5.64	18.23	26.85	17.72	20.38	27.65	50.73	32.64	15.29	7.55	10.63	9.51	13.23	13.51	22.06	16.59
Soluble REE content	La	3.43	26.26	88.93	39.54	60.13	120.42	47.88	76.14	5.22	74.17	17.09	32.16	22.93	73.62	51.30	49.28
	Ce	24.01	30.94	97.81	50.92	107.95	123.27	60.85	97.36	8.51	29.37	22.70	20.19	46.83	61.19	60.46	56.16
	Pr	0.70	4.97	19.13	8.27	14.97	25.33	7.66	15.99	1.66	18.14	4.74	8.18	5.78	16.15	10.51	10.81
	Nd	2.42	18.77	61.96	27.72	53.88	87.77	25.58	55.74	6.82	62.81	25.25	31.62	21.04	56.45	37.60	38.36
	Sm	0.36	2.39	9.41	4.05	9.19	12.80	3.88	8.62	1.40	12.75	5.35	6.50	3.65	9.31	6.21	6.39
	Eu	0.05	0.31	1.25	0.54	2.15	2.76	0.50	1.81	0.39	3.66	1.18	1.75	0.87	2.24	0.98	1.36
	Gd	0.35	1.64	6.41	2.80	6.48	8.53	2.99	6.00	1.43	13.20	5.41	6.68	2.75	7.79	4.94	5.16

Table 2 (Continued)

Tb	0.03	0.15	0.65	0.28	0.75	1.19	0.32	0.75	0.23	1.94	0.71	0.96	0.34	1.10	0.56	0.66
Dy	0.12	0.57	2.52	1.07	3.44	5.36	1.33	3.38	1.47	10.93	4.20	5.54	1.68	5.62	2.68	3.33
Ho	0.02	0.10	0.44	0.19	0.68	1.01	0.24	0.64	0.38	2.43	1.07	1.29	0.36	1.18	0.58	0.71
Er	0.05	0.25	1.15	0.48	1.88	2.71	0.64	1.74	1.19	7.02	3.38	3.86	1.04	3.33	1.72	2.03
Tm	0.01	0.03	0.14	0.06	0.26	0.37	0.08	0.24	0.19	0.99	0.54	0.57	0.15	0.46	0.25	0.29
Yb	0.03	0.16	0.76	0.32	1.65	2.20	0.46	1.44	1.19	5.91	3.49	3.53	0.96	2.76	1.57	1.76
Lu	0.01	0.02	0.11	0.05	0.25	0.33	0.06	0.21	0.19	0.94	0.56	0.56	0.15	0.43	0.25	0.27
Y	0.55	3.21	12.17	5.31	17.34	29.20	7.06	17.87	10.37	85.92	36.56	44.28	9.42	39.44	18.59	22.49
ΣREE	32.15	89.78	302.83	141.58	280.99	423.25	159.51	287.92	40.64	330.18	132.24	167.69	117.92	281.07	198.19	199.06
ΣCe	30.98	83.64	278.49	131.03	248.27	372.36	146.34	255.66	23.99	200.90	76.31	100.40	101.08	218.96	167.04	162.36
ΣY	1.17	6.14	24.34	10.55	32.72	50.90	13.17	32.26	16.64	129.28	55.93	67.29	16.84	62.11	31.15	36.70
ΣCe/ΣY	26.50	13.63	11.44	12.42	7.59	7.32	11.11	7.92	1.44	1.55	1.36	1.49	6.00	3.53	5.36	4.42
La(%)	16.39	44.39	69.99	43.59	95.43	84.44	26.86	59.51	11.55	66.31	14.88	35.49	53.27	70.42	36.63	51.40
Ce(%)	30.66	13.56	48.82	31.01	65.32	46.78	24.81	43.33	8.76	19.66	10.05	12.82	41.22	28.63	27.01	30.57
Pr(%)	19.60	52.44	79.24	50.43	103.99	75.84	21.48	57.47	16.74	73.78	17.38	39.73	62.15	71.80	36.20	53.33
Nd(%)	19.85	55.93	75.13	50.30	94.08	75.86	21.28	57.04	19.27	64.10	23.49	39.39	60.20	68.49	36.36	52.13
Sm(%)	16.44	50.51	85.01	50.65	98.49	76.54	26.40	63.50	22.38	71.20	27.92	45.02	61.67	70.98	41.50	56.38
Eu(%)	15.55	48.90	83.00	49.15	101.66	84.49	22.04	70.74	43.41	74.92	29.14	53.17	77.42	76.60	37.42	61.38
Gd(%)	13.56	38.66	80.28	44.17	98.35	72.25	30.16	63.57	31.20	69.41	30.61	48.55	60.08	66.66	41.64	55.03
Tb(%)	6.25	26.39	59.03	30.55	89.93	82.81	25.83	64.49	38.57	64.95	27.57	46.74	52.56	65.57	34.13	50.43
Dy(%)	3.41	15.54	42.63	20.53	79.92	71.82	22.15	56.99	45.21	59.20	28.69	45.66	45.19	56.97	30.31	44.50
Ho(%)	2.65	13.36	38.69	18.23	92.17	74.81	23.77	62.22	60.30	64.70	37.49	53.60	50.31	60.53	34.97	49.00
Er(%)	2.17	11.26	34.35	15.93	86.11	71.68	22.09	59.05	60.88	63.97	41.53	55.00	48.81	58.72	35.96	48.38
Tm(%)	1.89	8.43	27.98	12.77	83.09	70.77	20.47	58.08	55.48	63.81	47.51	56.76	45.54	57.48	37.53	47.88
Yb(%)	1.37	7.32	23.84	10.84	79.18	63.28	19.21	54.29	59.93	59.16	48.01	55.03	43.74	52.83	36.69	45.22
Lu(%)	1.45	7.71	25.23	11.47	83.05	70.75	20.09	59.24	70.01	68.82	54.64	63.51	48.28	60.63	41.15	51.08
Y(%)	2.50	15.05	34.03	17.19	80.95	70.28	23.19	57.39	59.38	72.23	44.77	60.92	46.29	65.08	37.73	51.79
ΣREE(%)	21.07	24.17	59.85	41.24	80.25	65.41	24.51	52.41	17.99	55.62	20.81	34.57	48.56	52.30	33.17	43.35
ΣCe(%)	26.36	24.91	62.36	43.68	79.73	64.74	24.54	51.72	12.32	49.41	15.30	27.38	48.62	49.85	32.51	41.96
ΣY(%)	3.34	17.21	41.01	24.34	84.42	70.84	24.15	58.61	53.54	69.11	40.82	56.87	48.18	63.26	37.24	50.79

The La(%) through ΣY(%) rows are grouped under the label "Soluble REE percentage".

Note: A, leached horizon; B, illuvium; C, Chorizon

4.3 Soluble REE

Generally the percentage of available REE (i.e. soluble REE) accounts for about 10%-20% of the gross of REE. But in Hainan Island, the percentage of available REE in granitic laterites is in the range of 17.99% (horizon A of laterites in middle mountain) to 80.25% (horizon A of laterites in low mountains-hills), and the average is 43.35%. The percentage of available light REE is from 12.32% (the leached horizon of laterites in middle mountain) to 79.73% (the leached horizon of laterites in low mountains-hills), and the average is 41.96%. The percentage of available heavy REE is from 3.34% (the leached horizon of laterites in tableland plain) to 84.42% (the leached horizon of laterites in low mountains-hills), and the average is 50.79%.

Among the different geomorphic units the average percentage of available REE is in the descending order of mountain-hills (52.41%) > tableland plain (41.24%) > middle mountain (34.57%). The change pattern of the average percentage of available light REE is the same as that of available REE, while the average percentage of available heavy REE is in the descending order of low mountain-hills (58.61%) > middle mountain (56.87%) > tableland plain (24.34%).

Among different genetic horizons the average percentage of available REE is in the descending order of illuvium B (52.30%) > leached horizon A (48.56%) > C horizon C (33.17%). In low mountain-hills, the order is leached horizon A (80.25%) > illuvium B (65.41%) > C horizon C (24.51%). In tableland plain, the order is C horizon C (59.85%) > illuvium B (24.17%) > leached horizon A (21.07%). So the change patterns of average percentage of available light REE and heavy REE are similar to that of available REE.

The above-mentioned results reflect that the gross available REE has a positive correlation with the gross of REE in one hand. On the other hand, they show that the high gross of available REE is related to the tropical climate with high temperature and moisture and to the strong chemical weathering, which results in the high maturity of soil and strong leaching. In addition REE transported chiefly in the form of soluble positive ions because of the acidic or weak acidic environment in laterites in Hainan Island.[9].

4.4 Change Patterns of the Average Gross of REE on the Horizontal Spatial Dimension

On the horizontal spatial dimension, the average gross of REE in granitic laterites in the Hainan Island is closely related to the special geomorphic structure of Hianan Island, which is unsymmetrical ring-layered structure with higher middle and lower surrounding areas. The average of REE in granitic laterites is in the order of low mountain-hills (549.34 µg/g) > middle mountain (485.06 µg/g) > tableland-coastal plain (343.33 µg/g) (Figure 3).

Average gross of light REE and available REE have the same change pattern as that of REE. Such as the average gross of light REE descends from low mountain-hills (494.29 µg/g), middle mountain (366.68 µg/g) to tableland-coastal plain (343.33 µg/g) (Figure 3). The average gross of available REE also descends from low mountain-hills (287.92 µg/g), middle mountain (167.69 µg/g) to tableland-coastal plain (141.58 µg/g), while the average gross of heavy REE is in the order of middle mountain (118.39 μg/g) > low mountain-hills (55.05 μg/g) > tableland-costal plain (43.34 μg/g).

The horizontal change patterns of average gross of total REE, light REE, available REE and heavy REE in granitic laterites are different. The possible reasons are that long term leaching by rain water caused light REE and soluble REE in middle maintain with steep terrain to transport to and enrich in low mountains-hills with low-lying land; but heavy REE trends to stay due to its relative stability. While in tableland-coastal plain, the REE is leached and transported to sea due to the relatively more strongly chemical weathering.

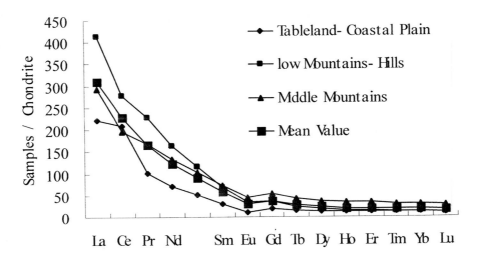

Figure 3: Distribution patterns of REE in granitic laterite in horizontal spatial dimension in Hainan Island.

4.5 Change Patterns of the Average Gross of REE on the Vertical Spatial Dimension

On the vertical spatial dimension the average gross of total REE decreases with the depth from bottom to top of the genetic horizon, such as Chorizon (597.48 μg/g) >illuvium (537.40 μg/g) > leached horizon (242.86 μg/g). Also, the average gross of light REE decreases with the depth in the order of Chorizon C (513.85 μg/g) > illuvium B (439.22 μg/g) > leached horizon A (207.90 μg/g). In the different geomorphic units, the average gross of total REE and light REE have the similar change patterns (Table 2, Figure 2 and Figure 3).

The average gross of heavy REE ($\sum Y$) gradually descends from illuvium B (98.18 μg/g), Chorizon C (83.64 μg/g) to leached horizon A (34.96 μg/g). Similar change patterns appear in central mountain area from illuvium B (187.06 μg/g), Chorizon C (137.01 μg/g) to leached horizon A (31.08 μg/g) and in low mountain-hills from illuvium B (71.84 μg/g), Chorizon C (54.55 μg/g) to leached horizonA (38.75 μg/g). But in the tableland plain the heavy REE is similar to light REE, decreasing from Chorizon C (59.35 μg/g), illuvium B (35.64 μg/g) to leached horizon A (35.03 μg/g).

The vertical change pattern of available REE is similar to that of heavy REE. The average gross of available REE ($\sum$REE) gradually declines from illuvium B (281.07 μg/g), Chorizon C (198.19 μg/g) to leached horizon A (117.92 μg/g). So does that in central mountain area from illuvium B (330.18 μg/g), Chorizon C (132.24 μg/g) to leached horizon A (40.64 μg/g) and in low mountain-hills from illuvium B (423.25 ug/g), Chorizon C (159.51 ug/g) to leached horizon A (280.99 μg/g). But in the tableland plain, the vertical change patterns of the average gross of available REE ($\sum$REE) and light REE ($\sum$Ce) are similar, i.e. they decrease from Chorizon C (302.83 μg/g), illuvium B (89.78 μg/g) to leached horizon A (32.15 μg/g) (Table 2).

The above-mentioned results reflect that REE in genetic horizons was influenced by mother material greatly. Tableland plain had the characteristics of strong leaching, which made REE enrich in Chorizon C and decrease in leached horizon A. While central mountains had the characteristics of relatively weak leaching, which made REE enrich in illuvium B and decrease in leached horizon A and Chorizon C.

4.6 Eu-Anomaly and Ce-Anomaly

The of value of δEu in granitic leterites is down from middle mountain (0.71), low mountain-hills (0.69) to tableland-coastal plain (0.46), which shows a weak to medium deficit of europium and negative Eu-anomaly.

Contrary to δEu, the standard value of δCe in granitic leterites is up from middle mountain (0.88), low mountain-hills (0.91) to tableland-coastal plain (1.39). The average δCe is near 1, which shows neutral Ce-anomaly. But the average of δCe is 1.39 in tableland plain (2.17 in its illuvium B and 2.32 in its leached horizon A), and 1.32 in leach horizon A in low mountain-hills, and 1.11 in leached horizon A. The above three values of δCe are more than 1.05, thus they show positive Ce-anomaly. With the soil depth, δCe changes from no Ce-anomaly to weak negative Ce-anomaly. The above characteristics of Ce-anomaly indicate that selective speciation (relative enrichment of cerium) occurred at the layer near soil surface (in leached horizon A) and in the tableland plain (at low altitude). The reason for the above mentioned results is that hydrolyzing reaction made Ce^{4+} separate from light REE, cerium changed from liquid phase to solid phase, and it stayed and accumulated at the top of laterites earlier.

5 CONCLUSION

According to the above results, the conclusions are as follows:

1) The average gross of total REE, light REE and available REE in granitic laterites in Hainan Island are higher than those in the crust. The characteristics of granitic laterites in enrichment of light REE, high content of available REE and positive correlation with total REE are related to the tropical climate with high temperature and moisture, strong chemical weathering and acidic to weak acidic soil environment.

2) On the horizontal spatial dimension, the average gross of total REE, light REE and available REE in granitic laterites in Hainan Island gradually descended from low mountains-hills, middle mountains to tableland plain, while the average gross of heavy REE in granitic laterites gradually decreased from middle mountains, low mountains-hills to tableland plain.
3) On the vertical spatial dimension, the average gross of REE and light REE in granitic laterites in Hainan Island gradually declined from bottom to top (Horizon C, Horizon B, Horizon A). While the average gross of heavy REE and available REE in granitic laterites in Hainan Island gradually decreased from Horizon C, Horizon B to Horizon A.
4) Granitic laterites in Hainan Island was with weak or medium deficit of europium, namely with negative Eu-anomaly. Cerium was enriched, namely positive Ce-anomaly in the leached horizon in granitic laterites of different geomorphic units and tableland plain with low altitude. But it was changed from positive Ce-anomaly, neutral Ce-anomaly to weak negative Ce-anomaly with the soil depth.

REFERENCES

[1] Hua, B. Qiang, L. Weihuang, Z. et al. (2003). Research progress in REE's use in agriculture, Journal of Hainan Normal University *Natural Science*, 16 (4): 77-81 (in Chinese).
[2] Gongpu, Z. Mingxun, C. & Guofeng, L. (1985). Agricultural geography of Hainan Island. Beijing: *Agriculture Press*, 1-6, 162-176 (in Chinese).
[3] Qingyi, Y. (1985). Landform conditions of Hainan Island. In: Guang Zhou Geography Institution. The tropical agriculture natural resources and layout of Hainan Island. Beijing: *Science Press*, 27-40 (in Chinese).
[4] Zhaoxuan, Z. & Xianzhong, Z. (1989). Physical geography of Hainan Island. Beijing: *Science Press*, 63-110, 272-312 (in Chinese).
[5] Huatang, C. (1989). Landform characteristics and unsing value of Hainan Island. In: Wen Changen, Yang Shigao, Fan Xinping, Li Yongxing. Hainan resource environment and research of roomage development. Haikou : Hainan People Press, 1989, 51-55 (in Chinese).
[6] Hua, B., Qiang, L., Longshi, Y. et al. (2002). Agricultural geomorphic features on Hainan Island. *GEOTECTONICA et METALLOGENIA*, 26(3): 326-330 (in Chinese).
[7] Shiying, S., Feng, H. & Jiantang, W. (1996). Hainan Travel. Beijing: *Chinese Personnel Press*, 1-115 (in Chinese).
[8] Roaldest, E. (1973). Rare earth elements in Quaternary clays of the Numedel area, Southern Norway, Lithos, 6 (4): 349-372.
[9] Zhonggang, W., Xueyuan, Y. & Zhenghua, Z. (1989). Geochemistry of REE. Beijing: *Science Press*, 76-93,321-341 (in Chinese).

INDEX

A

abatement, 52
abiotic, 136
abundance, 151
accounting, x, 181, 182
accuracy, viii, 16, 35, 36, 57
acetone, 95
acetonitrile, 95, 99, 103
acid, x, 32, 33, 35, 36, 40, 52, 53, 54, 95, 98, 99, 103, 107, 112, 114, 126, 131, 146, 147, 164, 166, 167, 168, 169, 170, 171, 179, 180, 181
acid mine drainage, 53
acidic, vii, 31, 32, 33, 34, 35, 37, 38, 39, 40, 41, 46, 51, 52, 53, 54, 55, 131, 187, 189
acidification, 170
acidity, 32, 33, 71
AD, 62, 158
adsorption, 177
Africa, 27, 159, 161
age(ing), 5, 6, 7, 23, 24, 26, 27, 28, 70, 134, 136, 144, 147, 148
agent, ix, 71, 93, 94, 95
aggregates, 39, 119, 121, 123, 124, 125, 127, 130, 134, 135, 136
aggregation, 131, 132
aggregation process, 131
agriculture, 144, 190
alanine, 126, 131
alcohol(s), ix, 93, 98, 99, 101, 110, 111, 112
aldehydes, ix, 93, 99
algae, 107, 108, 111, 132, 161
aliphatic compounds, 111
alkaline, 23, 94, 108, 109, 131
alkaline hydrolysis, 108, 109
alkane, 111
alkenes, ix, 93, 99, 112
alkynes, ix, 93, 98, 99, 112

ALS, 168, 170
alters, 126, 166
aluminium, 166
AMD, 32, 33, 37, 38, 41, 42
amendments, 177
amide, 128, 129, 137
amines, 104, 105
amino, 126, 131
amino acid(s), 126, 131
ammonium, 133
AMS, 147, 149
Amsterdam, 29, 54, 85, 138, 139, 140
amylase, 134
analytical techniques, 94, 106, 107
ANC, 171
animals, 183
anion, 62, 69
anoxia, 37
anthropogenic, 151, 161
antimony, 21
aquifers, 73, 85
arginine, 126, 131
arithmetic, 59
Armenia, 60, 85
aromatic compounds, 99, 101, 113
aromatic rings, 99, 110
aromatics, 95
ascorbic acid, 36
Asia, 84, 158
assessment, 52, 161, 179
association theory, 33
atmosphere, vii, 58, 161, 167
atomic emission spectrometry, 170
atomic force, 135
atoms, 81, 101, 128, 134
attention, 37, 50
Australia, v, x, 161, 163, 164, 165, 179
availability, 131, 164
averaging, 153

B

bacteria, 32, 122, 126, 137
banks, 41
behavior, 32, 43, 50, 61
Beijing, 57, 64, 81, 85, 88, 89, 90, 91, 143, 190
benzene, 99, 112, 114
binding, 125, 131, 137
bioavailability, x, 164, 168, 177, 179
bioconcentration, x, 163, 177, 178
biomarkers, 108, 111
biomass, x, 132, 136, 163, 170, 177
biomolecules, 108
biopolymers, 108, 113, 134
biosphere, 164
biotic, 136
bismuth, 21
Black Sea, 120
blocks, 10, 13, 15, 108, 148
bonding, 111
bonds, 95, 107, 108, 111, 129, 134
boundary surface, 59
breakdown, 132
British Columbia, 52, 54, 55
building blocks, 108
Burma, 87

C

Ca^{2+}, 42, 70, 73, 131, 132
cadmium, 160
calibration, 35, 40, 41
California, 32, 40, 52, 53, 54, 85, 86, 160
Canada, 23, 24, 25, 28, 32, 38, 52, 53, 54, 55
capillary, 164, 168, 173, 174
carbohydrate(s), 126, 128, 129, 131, 132, 133, 134, 135, 137
carbon, ix, 25, 62, 63, 66, 69, 81, 84, 86, 95, 99, 101, 103, 110, 111, 113, 123, 128, 143, 144, 146, 147, 148, 150, 151, 153, 154, 155, 156, 157, 158, 159, 160, 161, 168
carbon atoms, 101, 128
carbon dioxide, 62, 69, 81, 99, 161
carbon tetrachloride, 95
carboxylic acids, ix, 93, 94, 98, 99, 100, 101, 103, 106, 107, 114, 128
carrier, 81
case study, 61, 164, 177
catalyst, 95, 103, 109
cation, 62, 69
cell, 81, 108, 113, 126, 131, 134, 167, 168
cell membranes, 126
cellulose, 161
cement, 166
cerium, 189
certainty, 107
chemical composition, vii, ix, 33, 38, 39, 41, 64, 119, 137
chemical degradation, 106
chemical properties, 80
China, v, viii, ix, 57, 58, 59, 60, 61, 62, 63, 65, 66, 67, 70, 79, 81, 83, 84, 85, 87, 88, 89, 90, 91, 143, 144, 145, 147, 148, 149, 150, 151, 152, 154, 155, 158, 159, 160, 161, 181, 182
Chinese, 62, 80, 84, 85, 87, 88, 89, 90, 91, 143, 144, 147, 151, 159, 160, 183, 190
chloride, 90, 111
chloroform, 95
chlorophyll, 132, 133
chromatography, 126, 127
circulation, 74, 135
classes, 122, 128, 137
classification, vii, 1, 3, 11, 17, 23
cleavage, 111, 114, 134, 135, 137
climate change, 8, 10, 23, 144, 160, 161
climatic change, 161
climatic factors, 144
closure, 164
clusters, 59
CO_2, 42, 53, 66, 81, 89, 146, 156, 157, 158
coal, 106, 107
cobalt, 6
codes, 36, 41, 43, 51
combustion, 146
communication, 51
community, 127, 131, 135, 136
compilation, 33, 40
complexity, 51, 71, 91, 94
complications, x, 144, 159
components, vii, 9, 59, 61, 63, 64, 77, 81, 82, 122, 124, 128, 129, 134, 136, 137
composition, vii, ix, 1, 2, 6, 9, 10, 11, 12, 14, 15, 21, 22, 26, 32, 33, 37, 38, 39, 40, 41, 51, 86, 93, 108, 110, 113, 119, 120, 122, 123, 124, 125, 126, 128, 129, 134, 135, 136, 137, 144, 146, 161, 170
compounds, ix, 95, 99, 101, 103, 107, 108, 111, 113, 114, 119, 127, 128, 129, 134, 137
concentrates, 10, 17, 35
concentration, vii, x, 10, 21, 31, 32, 36, 37, 40, 41, 42, 43, 44, 46, 48, 49, 51, 53, 58, 59, 63, 64, 65, 66, 68, 73, 74, 78, 81, 83, 87, 108, 125, 126, 131, 132, 136, 150, 163, 167, 173, 177, 178, 184
concentration ratios, 78
conceptual model, 78
condensation, 134

conductivity, 35, 37, 79, 80, 81, 91
confidence, ix, 59, 93
confidence interval, 59
conifer, 156
consensus, 131
consolidation, 135
constraints, 23, 86, 90
construction, x, 62, 163, 164, 167
consumption, 136, 157
contaminants, 164
contamination, 5, 149, 174
control, 15, 33, 49, 81, 168, 170
cooling, x, 49, 144, 145, 158, 159
copper, 6, 20, 21, 160
correlation(s), x, 27, 58, 59, 60, 181, 187, 189
costs, 164
coupling, 90
coverage, 154, 157
covering, 33, 121, 167
crude oil, 106, 107
crystalline, 170
crystals, vii, 31, 33, 34, 35, 36, 39, 46, 51
cyanobacteria, 114, 122
cycles, vii, 154, 158
cycling, 137
cyclopentadiene, 111

derivatives, 99
desiccation, 173
detection, 35, 36, 37, 38, 63, 146
deviation, 59
diamond, 9, 19
differentiation, ix, 27, 93, 114
diffraction, 170
diffusion, 80, 82
digestion, 36, 147
disaster, 61
discipline, 60
discontinuity, 81
discriminant analysis, 59, 87
dispersion, vii, 1, 2, 3, 9, 12, 14, 17, 19, 20, 22, 24
dissolved oxygen, 37
distilled water, 36, 168, 170, 171
distribution, ix, 24, 36, 62, 90, 108, 119, 120, 151, 153, 161
division, 11, 17
double bonds, 95, 108
drainage, 32, 33, 34, 37, 38, 40, 41, 52, 53, 54, 153, 155, 157, 160, 164, 179, 180
drought, x, 144, 159, 161, 177
drying, 64, 183
duration, 59, 60, 70, 81, 125, 135, 165
dykes, 6

D

data analysis, 58, 71
database, 2, 18, 36, 41, 42, 43, 49, 50, 51, 52, 53
dating, ix, 7, 143, 147, 159
Davis approach, viii, 31, 50
Davis equation, viii, 31, 42, 43, 46, 49, 50, 51
Dead Sea, 108
deaths, viii, 57, 83
Debye, 32, 33, 36, 41, 42, 43, 51, 53
decoding, ix, 119
deficit, 184, 189, 190
definition, 41, 42
deformation, 5, 10, 11, 59
degradation, viii, ix, 93, 94, 99, 106, 107, 108, 109, 111, 112, 113, 114, 119, 124, 125, 126, 131, 132, 134, 135, 136, 137
degradation process, ix, 119, 124, 134, 135
dehydration, 81, 82, 91, 167
density, 10, 16, 32, 60, 99, 120, 135, 166
Department of Energy, 159
deposition, vii, 1, 2, 5, 8, 10, 11, 12, 13, 14, 15, 17, 18, 22, 23, 136, 153
deposits, vii, 1, 2, 3, 6, 7, 8, 9, 11, 13, 15, 23, 26, 27, 28, 40, 53, 81, 144, 159, 161, 179
depression, 15

E

earth, xi, 6, 58, 59, 71, 78, 79, 80, 81, 82, 83, 182, 190
earthquake, viii, 57, 58, 59, 60, 61, 62, 63, 64, 66, 67, 68, 69, 70, 71, 72, 73, 74, 78, 79, 80, 81, 82, 83, 84, 85, 86, 87, 88, 89, 90, 91
East Asia, 160, 161
ecology, 53
ecosystem, 137
education, 85
effluent, 166, 167
eggs, 136
electric conductivity, 35, 37
electrical conductivity, 80
electrical properties, 28
electrodes, 40
electrolyte(s), 32, 36, 42, 43, 53, 54
electron, 10, 17, 99, 101
electron density, 99
emission, 85, 170
employees, 159
encapsulation, 134
energy, vii, 58, 59, 78, 79, 81, 150, 153
entrapment, 125

environment, vii, ix, x, 1, 2, 3, 8, 9, 11, 14, 15, 17, 22, 26, 28, 53, 55, 136, 144, 150, 154, 155, 156, 158, 159, 164, 181, 187, 189, 190
environmental conditions, 35, 53, 120, 137
environmental impact, 179
enzymatic activity, 134
enzymes, 131, 134
EPA, 52
equilibrium, viii, 31, 33, 46, 49, 51, 52, 58, 78, 144
equipment, 2, 16
erosion, 7, 9, 10, 14, 15, 17, 22, 164, 165, 177
ester bonds, 111
ester(s), ix, 93, 94, 98, 101, 103, 105, 107, 109, 110, 111, 112, 114, 128, 129, 137
ether(s), ix, 93, 95, 98, 99, 101, 103, 105, 107, 110, 111, 112, 113
ethyl acetate, 95
Eurasia, 27, 28, 70
Europe, 10, 27, 155, 158
European Union, 23
europium, 189, 190
eutrophication, 138
evaporation, 32, 35, 37, 39, 40, 46, 49, 51, 145, 146, 158, 165
evidence, x, 2, 4, 5, 87, 90, 91, 128, 144, 155, 156, 157, 158, 159, 161, 163, 164
evolution, 23, 26, 29, 90, 132, 160, 164
exclusion, 126, 127
excretion, 132
exercise, vii, 1
exposure, 167
extraction, 94, 108, 168

F

fabric, 5, 13, 22
facies, 4, 5, 11
factor analysis, 59, 71
failure, viii, 57, 83
family, 107
fatty acids, 108, 113, 126, 160
Fennoscandian, v, vii, 1, 3, 6, 7, 24, 25, 29
fibrils, 135
filters, 35
filtration, 166
Finland, vii, 1, 2, 3, 4, 6, 7, 8, 9, 12, 13, 14, 16, 17, 20, 21, 22, 23, 24, 25, 26, 27, 28, 29
fires, 154
fish(ing), 135, 136
fisheries, 136, 137
flood, 165, 166
flotation, 166
fluctuations, ix, 85, 144, 149, 159

fluid, viii, 58, 59, 60, 61, 62, 63, 71, 72, 73, 77, 78, 79, 81, 83, 84, 85, 86, 88, 89
fluid transport, 81
fluoride, 161
foams, 120
focusing, ix, 143
food, 136, 177
forecasting, 60, 85, 137
forest fires, 154
fossil, ix, 24, 93, 107, 108, 114, 143
fractures, 70, 74, 82
France, 23, 123
fructose, 126
FTIR, ix, 108, 119, 127, 129, 130, 134
FTIR spectroscopy, 127
fuel, 107
fuel type, 107

G

gases, 58, 61, 81, 84
gel, 123, 131, 135
generation, viii, 23, 32, 57, 58, 62, 79, 80, 81, 164, 170
geochemical, vii, viii, ix, x, xi, 1, 2, 3, 5, 7, 9, 12, 14, 16, 20, 23, 25, 26, 27, 28, 29, 32, 33, 36, 40, 41, 43, 50, 52, 54, 57, 58, 59, 60, 61, 62, 63, 64, 70, 71, 72, 73, 77, 79, 81, 83, 84, 87, 89, 92, 143, 144, 149, 153, 158, 159, 172, 179, 182, 184
geography, 143, 190
geology, vii, 1, 2, 24, 28, 29
Germany, 81, 91, 160, 163
gestation, viii, 58
GIS, 3, 9, 24, 26, 27
glacial deposits, vii, 1, 2, 7, 13, 26
glaciations, vii, 1, 3, 8, 12, 15
glaciers, vii, 1, 7, 10
glass, 35, 40, 41, 63, 147
global warming, 156
glucose, 126
glutamic acid, 126, 131
glycine, 126, 131
glycoside, 137
goals, viii, 57
gold, 1, 6, 10, 12, 17, 19, 20, 21, 22, 24, 26, 27
government, 164
grains, 21, 39, 78
graphite, 166
grass(es), 154, 156, 165, 166, 169, 170, 174, 176, 177, 178
grasslands, 165, 166
gravity, 11, 33, 63
grazing, 135, 166, 177

Greenland, 10, 160
grids, 2, 9, 22
groundwater, 7, 21, 22, 58, 61, 67, 72, 74, 77, 84, 87, 88, 89, 90
groups, ix, 5, 6, 60, 93, 94, 99, 109, 112, 113, 114, 122, 128, 129, 130, 137
growth, 10, 32, 82, 108, 126, 135, 164
Guangdong, 88, 183
Guangzhou, 143

H

Hainan Island, v, x, xi, 181, 182, 183, 184, 185, 187, 188, 189, 190
half-life, 82
health, 136, 179
heat(ing), 58, 79, 80, 91
heavy metals, 125, 136, 154, 155
height, 11, 12, 13
helium, 62, 81, 85
hematite, 21
hemisphere, 144, 158, 159
heterogeneity, 90
histidine, 126, 131
holocene, ix, 144, 148, 150, 154, 156, 157, 158, 159, 160, 161
host, 6, 33, 79
house, 23, 25
humic substances, 134
humidity, x, 35, 181
hybrid, 52
hydrocarbons, 99, 107, 126
hydrochemical, viii, 38, 52, 57, 62, 63, 69, 71, 73, 77, 83, 87, 88, 90
hydrogen, 36, 37, 40, 81, 129, 171
hydrogeochemical, vii, 31, 51, 67, 88, 89, 91
hydrolysis, 94, 103, 108, 109, 112, 113, 132, 134, 171
hydrosphere, vii
hydroxide, 112, 170
hydroxyl, 130, 160
hydroxyl groups, 130
hypothesis, 132

I

ice caps, 25
ice lobation, vii, 1, 8
ice-flow, vii, 1, 2, 8, 9, 11, 15, 22
identification, vii, 1, 2, 3, 9, 10, 22, 23, 39, 59, 71, 107
images, 9

impurities, 39, 41, 50
inclusion, 50
India, 70, 85, 86, 89
Indian Ocean, 160
indication, 2, 21, 106, 107, 109, 134, 150
indicators, 20, 23, 27
indices, 33, 36, 41, 51
indirect effect, 136
industry, 33
insertion, 103
insight, viii, 93, 129
inspection, 173, 174
instability, 87
instruments, 35, 62
integration, 26
integrity, 164
intensity, 2
interaction(s), vii, viii, 5, 31, 32, 33, 36, 41, 42, 43, 50, 51, 52, 53, 54, 58, 72, 77, 83, 124, 125, 131, 134, 135, 164
interface, 175
internal consistency, 50
interpretation, vii, 1, 9, 28
interval, 13, 15, 59, 150, 156, 157
intervention, 59
intrusions, 3, 5, 6, 183
inversion, 25
ions, viii, 32, 41, 42, 49, 51, 54, 58, 78, 84, 124, 131, 132, 134, 171, 187
IR spectra, 130
IR spectroscopy, 129
iron, 6, 7, 8, 36, 37, 38, 39, 124, 157, 158, 171
Iron Age, x, 144, 158, 159
isolation, 37, 132, 164
isoleucine, 126, 131
isotope(s), ix, 84, 86, 91, 124, 144, 146, 151, 152, 153, 155, 156, 157, 158, 159, 160, 161
Isotopic, 84, 89, 146
Italy, 24, 32, 38, 40, 53, 86, 87, 92

J

Japan, 60, 86, 87, 89

K

K^+, 42
KBr, 130
kerogen, viii, 93, 94, 106, 107, 108, 109, 111, 112, 113, 114, 115, 147
ketones, 95, 101, 103, 108
King, 61, 82, 85, 86

Kobe, 86

L

lactones, 103
Lake Gucheng, v, ix, 143, 144, 145, 146, 147, 148, 149, 150, 151, 152, 153, 154, 155, 156, 157
lakes, 9, 38, 55, 144, 156
land, x, 139, 166, 181, 182, 183, 188
land use, 166
lanthanum, 184
Lapland, vii, 1, 3, 4, 5, 6, 7, 8, 9, 12, 14, 16, 17, 18, 19, 20, 21, 23, 24, 25, 26, 27, 28
larvae, 122
Last Glacial Maximum, ix, 25, 144, 145, 159
Late Quaternary, ix, 28, 143
leachate, 32, 34, 37, 39, 166
leaching, 168, 170, 171, 187, 188, 189
lead, 41, 108, 135, 136, 160, 171
learning, 59
lending, 183
leucine, 126, 131
likelihood, 164
limitation, 41, 42, 60, 132
linkage, 134
lipase, 127, 134
lipids, 109, 111, 126, 134, 135, 137
lipopolysaccharides, 127
liquid phase, 189
liquids, 10
literature, 16, 37, 38, 95, 183
Little Ice Age, 158
location, ix, 8, 12, 60, 143, 145, 173, 174
London, 137
long distance, 82
lying, 188
lysine, 126
lysis, 131

M

machine learning, 59
macroaggregates, ix, 119, 120, 122, 123, 124, 125, 126, 127, 128, 129, 130, 131, 132, 133, 134, 135, 136, 137
macrogels, ix, 119, 131, 137
macromolecular networks, 111
macromolecules, 94, 106, 109, 131
magnet, 147
magnetite, 7, 20, 21, 166, 170
management, 179
mantle, 4, 24, 79, 80, 81, 91

mapping, 1, 2, 25, 26, 27
marine environment, 136
mass spectrometry, 170
Massachusetts, 54
matrix, 12, 110, 112, 131, 134, 135, 136
maturation, 134, 135
measurement, 37, 41, 58, 62, 89, 146, 147
measures, 139
media, 168
median, 158, 170, 171, 172, 173
melanterite, vii, 31, 32, 33, 34, 35, 36, 39, 40, 41, 43, 45, 46, 47, 48, 49, 50, 51, 53
melt(ing), 3, 8, 11, 17, 19, 80, 81, 91, 156
membranes, 126
memory, 160
Merck, 35
mercury, 62, 63, 64, 65, 68, 81, 83, 89
metal content, 32, 37, 125, 152
metal ions, 171
metalloids, x, 164, 168, 175, 177, 178
metals, x, 6, 10, 26, 32, 33, 37, 125, 135, 136, 151, 154, 155, 156, 158, 161, 163, 164, 168, 173, 174, 175, 177, 178
methionine, 131
methylene, 95, 105
methylene group, 95
Mexico, 86
Mg^{2+}, 39, 42, 73
microbial, 53
microbial community, 131, 135
microgels, 131
micrograms, 152
microorganisms, 32, 35, 49
microscope, 10, 17
microscopy, 123, 135
migration, 58, 72, 73, 79, 81, 82, 83, 90, 164
mineralized, x, 12, 14, 16, 17, 18, 19, 22, 163, 168, 169, 173, 175, 177, 178
minerals, 1, 7, 10, 15, 20, 21, 22, 25, 26, 28, 33, 35, 41, 46, 51, 54, 82, 91, 124, 134, 143, 144, 150, 151, 153, 155, 156, 157, 158, 159, 166, 170
mining, 32, 33, 41, 52, 53, 166
Miocene, 160
missions, 161
mixing, viii, 7, 58, 59, 73, 87, 94, 135
modeling, vii, 31, 32, 33, 40, 52, 53
models, viii, 23, 24, 31, 33, 42, 46, 50, 80
moieties, ix, 93
moisture, 177, 183, 187, 189
molecular weight, 107, 113, 126, 127, 129, 134, 135
molecules, 39, 101, 106, 114, 131
mollusks, 147
molybdenum, 24

Mongolia, 57, 161
monosaccharide(s), 126, 133, 135, 137
monsoon, ix, 143, 144, 145, 155, 156, 157, 158, 159, 161
morning, 37, 46, 51
morphology, 1, 2, 9, 10, 11, 13, 17, 21, 22, 27
Moscow, 23
motion, 135
mountains, 182, 183, 185, 187, 188, 189, 190
movement, 7, 10, 12, 13, 173

N

Na^+, 42
NaCl, 74
National Science Foundation, 181
native plant, x, 163, 167, 177, 178
natural resources, 190
Nd, 185, 186
neoglaciation, x, 144, 154, 158, 159
Netherlands, 25
network, viii, 57, 60, 61, 62, 63, 70, 83, 88, 111, 123, 135
neural networks, 87
New York, 84, 115, 138, 141, 143, 179
nickel, 6, 111, 114
nitrate, 133
nitric acid, 147
nitrogen, 105, 133, 134
NMR, ix, 106, 108, 119, 127, 128, 129, 130, 134
N-N, 70, 71
nodules, 172
North America, 7, 158, 159
North Atlantic, 155
North Sea, 120, 141
Norway, 28, 190

O

observations, 8, 9, 23, 24, 41, 46, 51, 81, 89, 131, 161
oil(s), viii, 64, 91, 93, 94, 106, 107, 109, 111
oil sands, viii, 93
oil shale, viii, 93, 109
optimization, 114
ores, 144, 166, 170
organic compounds, ix, 95, 114, 119, 127
organic matter, ix, 93, 94, 107, 108, 109, 114, 119, 123, 126, 128, 131, 132, 134, 135, 137
organic solvents, 94, 95
organization, 135
orientation, 2, 12

oscillation, 7, 66
Ottawa, 52
oxidants, ix, 93, 95
oxidation, viii, 32, 37, 40, 41, 49, 51, 52, 54, 93, 94, 95, 99, 101, 103, 106, 107, 108, 109, 111, 112, 113, 114, 116, 167, 171
oxidation products, 94, 103, 106, 107, 109
oxygen, ix, 32, 37, 86, 103, 111, 114, 128, 136, 143, 144, 146, 151, 153, 154, 155, 156, 158, 159, 160
oxygen consumption, 136
ozone, ix, 93
ozonolysis, 99

P

Pacific, 161
pairing, 33, 42
parameter, 19, 59
Paris, 115
particles, 122, 124, 125, 129, 132, 135, 136, 137, 154, 167
particulate matter, ix, 119, 125
partition, 129
passive, 11, 22
pathogens, 136
pathways, 86
peat, 12, 161
performance, x, 163, 164, 167, 178
petroleum, 143
pH, vii, 31, 32, 33, 35, 36, 37, 40, 41, 43, 44, 45, 46, 47, 48, 49, 50, 51, 52, 54, 69, 71, 94, 131, 146, 166, 167, 168, 170, 171, 172, 173
phase boundaries, 132
phenylalanine, 126, 131
phosphate, 132, 133, 177
phosphorus, 132
photochemical degradation, 134, 137
photochemical transformations, 134, 137
photooxidation, 133
photopolymerization, 133
photosynthesis, 144
physical chemistry, 54
physico-chemical characteristics, 32
physics, 81
phytoplankton, 122, 126, 132, 134, 135, 137
Pitzer approach, viii, 31, 42, 43, 44, 49, 50, 52
Pitzer model, viii, 31, 40, 42, 43, 46, 49, 50, 51
planets, vii
plankton, 131, 136, 137
planning, 8, 9
plants, x, xi, 158, 163, 164, 168, 177, 178, 182, 183
plasma, 170
platinum, 6

Pleistocene, ix, 24, 144, 145, 148, 150, 153, 155, 156, 157, 159, 160
pollen, 122, 153, 154, 155, 156, 157, 158
pollution, 41
polycyclic aromatic compounds, 99
polyethylene, 35
polymer, 106, 116, 131, 135
polymerization, 134
polysaccharide, 132, 134, 135
pools, vii, 31, 32, 34, 37, 39, 40, 46
poor, 95, 99
population, 61
positive correlation, x, 181, 187, 189
potassium, 112
precipitation, viii, ix, 8, 31, 34, 35, 41, 46, 49, 51, 77, 144, 145, 146, 155, 156, 157, 158, 159, 161, 177
prediction, viii, 53, 57, 59, 60, 61, 80, 83, 85, 86, 87, 89
preference, 114
pressure, viii, 8, 11, 12, 13, 58, 59, 72, 78, 81, 91, 135
probability, 70
probe, 35
producers, 131
production, 32, 33, 131, 134, 135, 136, 137, 155, 156, 157, 158, 167, 171
program, 36, 51, 53, 54, 59
protein(s), 124, 126, 129, 134, 135, 137
proteinase, 134
protons, 32
pulse, 157, 158
pyrite, vii, 14, 21, 31, 32, 33, 34, 37, 39, 40, 41, 46, 51, 53, 146, 166
pyrolysis, 94, 109

Q

quartz, 6, 14, 21, 34, 124, 125, 129, 143, 150, 154, 155, 156, 157, 159, 166, 171, 172

R

race, ix, 143, 144, 156, 159, 172, 173
radar, 28
radiation, 36, 85, 86, 155
radius, 60, 71
radon, 58, 59, 60, 62, 63, 64, 65, 66, 67, 74, 81, 85, 86, 87, 88
rain, 135, 158, 167, 169, 188
rainfall, 58, 69, 165
Raman, 135

Raman spectroscopy, 135
range, x, 5, 9, 15, 36, 40, 41, 42, 43, 46, 48, 49, 80, 81, 107, 123, 128, 129, 130, 153, 165, 167, 181, 187
rare earth elements, xi, 182
reaction temperature, 106, 107
reaction time, 98
reactivity, 95, 99, 113, 114, 126, 131
reagents, 103
recession, 157
reconstruction, viii, 2, 93, 94
recovery, 33, 81
redox, 37, 53
reduction, 37, 94, 111, 132, 133
refractory, 129, 134
regional, 5, 9, 16, 17, 26, 27, 29, 60, 144, 154
regolith, vii, 1, 7, 8, 22, 28
regression, 59, 64, 147, 149
regression analysis, 59
regression method, 64
regulation, 62
rehabilitation, 164, 175, 177
relationship(s), viii, 57, 59, 60, 61, 69, 71, 72, 80, 86, 90, 144, 148
relaxation, 81
reliability, 52, 70
reparation, 39, 87, 125
residues, 126
resins, 116
resistance, 69, 135
resources, 81, 190
retention, 132, 169
rheology, 80, 90
rhizosphere, 177
ribose, 126
rings, 99, 100, 110
risk, 136
rolling, 64
room temperature, 94, 95, 168
Royal Society, 25
runoff, 33
Russia, 7, 84
ruthenium, viii, 93, 94, 95, 98, 99, 101, 103, 104, 105, 106, 107, 108, 109, 111, 112, 113, 114

S

safety, 94
salinity, 131, 149, 157
salt(s), 32, 35, 40, 41, 94, 156, 164, 167, 175
salt formation, 32, 40
sample, ix, 10, 15, 21, 36, 39, 107, 128, 129, 130, 134, 143, 146, 149, 153, 168, 170, 183

Index

sampling, vii, 1, 2, 8, 9, 15, 16, 20, 22, 31, 32, 35, 37, 46, 49, 51, 67, 125, 159, 183
San Telmo, v, vii, 31, 32, 33, 37, 38, 39, 40, 41, 42, 46, 49, 51
satellite, 9
saturation, 33, 36, 41, 43, 46, 49, 50, 51
scaling, 40
scatter plot, 60
sea level, 24, 145, 160
search, 27, 85, 170
seasonal variations, 59
seawater, ix, 36, 119, 124, 125, 126, 131, 132, 134, 136, 137
SEC, 126, 127
sediment(s), x, 2, 5, 10, 11, 12, 13, 15, 94, 108, 124, 125, 132, 144, 146, 147, 148, 149, 150, 151, 152, 153, 154, 155, 156, 158, 159, 160, 161, 165, 166
sedimentation, 3, 9, 23, 135, 136, 147, 150, 153, 160
seed, 167
selecting, 36, 49, 70, 177
selectivity, 94, 95, 114
separation, 8, 10
series, 4, 59, 65, 68, 71, 77, 86, 108, 109, 112, 161, 166
serine, 126, 131
shape, 17, 21, 120, 182
shear, 20, 120, 132
shock, 23, 60, 61, 73
sign, 50, 51
signals, ix, x, 58, 61, 64, 71, 87, 128, 134, 143, 144, 153, 154, 159, 160
silica, 124
silicate, 21, 129
silicon, 125, 128
SiO2, 39
sites, 32, 37, 41, 55, 59, 70, 71, 86, 125, 164, 168, 176, 183
sludge, 34, 39
sodium, 94, 166
software, 3, 33, 36, 50
soil, x, 9, 20, 58, 59, 64, 65, 66, 69, 85, 86, 87, 106, 148, 154, 163, 164, 166, 167, 168, 169, 170, 172, 173, 174, 175, 177, 178, 179, 181, 183, 187, 189, 190
solid phase, 189
solid state, 106, 128
solubility, vii, 31, 33, 36, 41, 43, 45, 46, 47, 48, 49, 54, 168
solvent(s), 94, 95, 108
South China Sea, 182
Spain, v, vii, 26, 28, 31, 32, 53, 55, 139
speciation, 37, 43, 52, 53, 54, 189

species, x, 32, 36, 42, 46, 49, 50, 95, 108, 122, 126, 131, 134, 135, 156, 163, 165, 166, 167, 174, 175, 177, 178
spectroscopy, 127, 128, 129, 135
spectrum, 46, 128, 129, 130
speed, 81
spore, 154, 155, 156
stability, 125, 135, 164, 188
stabilization, 177
stages, ix, 6, 62, 109, 126, 135, 144, 159
standard deviation, 59
standards, 35, 40
stars, 79, 80
statistics, 61
steel, 168
steroids, 99, 111
storage, 91, 126, 164, 166, 178
strain, 4, 59, 63, 71, 73, 82
strategies, 9, 167
stratification, 132
strength, viii, 32, 33, 42, 43, 49, 50, 51, 52, 79, 153
stress, 11, 58, 73, 77, 82, 83, 132
stress fields, 11
striae, 9, 22
structural characteristics, 90
structural transformations, 134
substitution, 39, 128
substrates, 135, 175
sulfate, 166, 167
sulfur, 108, 111, 146, 148, 150, 167, 170
sulfuric acid, 166
sulphur, 126, 135
summer, 46, 64, 120, 126, 128, 132, 134, 135, 136, 144, 145, 155, 160, 165
Sun, 91
superfund, 54
surface area, 74
surface layer, 119, 121, 123
sustainability, 164
Sweden, 24
swelling, 166, 172
synthesis, 87
systems, 3, 24, 32, 38, 52, 90, 164

T

Taiwan, 64, 65, 67, 69, 70, 86, 88, 90, 182
talc, 166, 170
tea, 161
technology, 24
temperature, viii, x, 4, 35, 37, 40, 43, 46, 48, 51, 54, 58, 63, 78, 81, 82, 91, 94, 95, 107, 132, 144, 145, 146, 147, 155, 158, 168, 181, 183, 187, 189

tension, 13, 15
tetrahydrofuran, 103
tetroxide, viii, 93, 94, 95, 98, 99, 101, 103, 104, 105, 106, 107, 108, 109, 111, 112, 113, 114, 116
textbooks, 43
theory, 33, 36, 43, 53, 61, 79, 131
thermodynamic, 36, 42, 46, 49, 50, 51, 52, 53
thermodynamic properties, 50
threonine, 126, 131
thresholds, 59
time, vii, viii, 9, 15, 17, 23, 37, 58, 59, 60, 65, 66, 71, 77, 78, 82, 86, 87, 95, 107, 108, 120, 131, 132, 136, 137, 149, 156, 158, 161, 183
time series, 59, 65, 77, 86, 161
timing, 24, 26, 27
tissue, x, 164, 178, 179
tourism, 136, 137
trace elements, 35, 36, 124, 125, 151, 156, 168, 170, 172, 173
transformation(s), ix, 101, 119, 131, 132, 134, 135, 136, 137
transition, 5, 154, 155
translocation, x, 163, 177, 178
transmission, 73
transpiration, 177
transport, vii, 1, 3, 9, 11, 12, 14, 15, 16, 17, 18, 19, 20, 21, 22, 33, 35, 54, 58, 82, 90, 136, 165, 188
transportation, 2, 10, 11, 12, 14, 17, 19, 22
trees, 59, 166
trend, 46, 71, 113, 131, 153
trial, x, 163, 167, 168, 169, 170, 173, 174, 175, 176, 177, 178
turbulence, 120, 133
turbulent, 132, 135
turbulent mixing, 135
Turkey, 84, 86
tyrosine, 126, 131

U

UK, 137
ultrasound, 108
uniform, 126
United States, 159
universities, 23
uranium, 6, 20, 164
USSR, 84

V

validity, 33

valine, 126, 131
values, ix, x, 19, 32, 35, 37, 45, 46, 49, 51, 65, 66, 73, 123, 124, 125, 131, 134, 136, 143, 144, 151, 152, 153, 155, 156, 157, 158, 159, 163, 170, 171, 172, 173, 175, 177, 181, 184, 189
variability, 27, 86, 136, 161
variable(s), vii, 1, 8, 11, 12, 15, 17, 59, 70, 71, 165, 166, 167, 172
variation, 2, 9, 11, 15, 58, 59, 61, 64, 65, 69, 71, 74, 77, 85, 88, 132, 154
varimax rotation, 59
vegetation, 12, 154, 157, 164, 166, 168, 169, 170, 177
vein, 14
velocity, 70, 79, 80, 81, 82, 90, 91
village, 14
Virginia, 54
volcanic activity, 90

W

war, 8
Washington, 87, 90, 139, 140, 161
wastewaters, 52
water evaporation, 51
watershed, 55
web, 136
wells, 62, 63, 64, 67, 88
wetting, 171
wind, 33, 135, 154, 155
winter, 46, 64, 144, 165
wood, 154, 155, 156, 157, 158
workers, 40, 107, 109, 111

X

X-ray diffraction (XRD), 36, 39, 130, 167, 170, 172

Y

yield, 32, 35, 99, 101, 103, 106, 112
ytterbium, 184

Z

zinc, 160
zooplankton, 122, 127, 135